Contemporary ELT Strategies in Engineering Pedagogy

W0113233

This book explores innovative pedagogical practices and teaching and learning strategies in the engineering curriculum for empowered learning.

It highlights the exigency for developing specific skill sets among students that meet the current market recruitment needs. The authors present a detailed framework for fostering a higher level of competence in students especially in their communication skills, their knowledge of media and technology tools, and their leadership skills. The book offers examples of new and effective teaching strategies including cognitive, metacognitive, and socio-affective strategies which align well with the existing and evolving technical curriculum.

The book will be of interest to teachers, students, and researchers of education, engineering, and higher education. It will also be useful for English language teachers, educators, and curriculum developers.

S. Mekala is an Associate Professor and Head in the Department of Humanities and Social Sciences at National Institute of Technology, Tiruchirappalli, India.

Geetha R is a doctoral candidate who completed her PhD research focusing on the augmentation of learners' writing skill, in the Department of Humanities and Social Sciences at National Institute of Technology, Tiruchirappalli, India.

Contemporary ELT Strategies in Engineering Pedagogy

Theory and Practice

Edited by
S. Mekala and Geetha R

Routledge
Taylor & Francis Group

LONDON AND NEW YORK

First published 2022
by Routledge
4 Park Square, Milton Park, Abingdon, Oxon OX14 4RN

and by Routledge
605 Third Avenue, New York, NY 10158

Routledge is an imprint of the Taylor & Francis Group, an informa business

© 2022 selection and editorial matter, S. Mekala and Geetha R; individual chapters, the contributors

The right of S. Mekala and Geetha R to be identified as the authors of the editorial material, and of the authors for their individual chapters, has been asserted in accordance with sections 77 and 78 of the Copyright, Designs and Patents Act 1988.

British Library Cataloguing-in-Publication Data
A catalogue record for this book is available from the British Library

Library of Congress Cataloging-in-Publication Data
A catalog record has been requested for this book

ISBN: 978-0-367-72517-4 (hbk)
ISBN: 978-1-032-21463-4 (pbk)
ISBN: 978-1-003-26852-9 (ebk)

DOI: 10.4324/9781003268529

Typeset in Sabon
by SPi Technologies India Pvt Ltd (Straive)

Access the [companion website/Support Material]: [insert comp website/ Support Material URL]

Contents

List of Figures   viii
List of Tables   ix
Preface   xi
Acknowledgments   xiii
List of Contributors   xiv

PART I
Developing Competency and Skills   1

1 Need for Remodelling the Engineering Curriculum   3
  C. HARISHREE

2 Integrated Practice of Learning Strategies in the
  Engineering Curriculum   20
  GEETHA R

PART II
Communication Skills   39

3 Bridging the Skills Gap in Technical Education   41
  C. SHANMUGA PRIYA

4 Task-Based Language Teaching to Foster
  Communication Skills   61
  N. CHITRA

PART III
Listening, Speaking, Reading and Writing Skills   81

5 Teaching LSRW Skills through the Test, Teach, Test (TTT)
   Method   83
   P. HILTRUD DAVE EVE

6 Integrating the Sub-Skills of LSRW for ESL Learners   95
   R. VASANTHAN AND R. NANDHINI

PART IV
English for Specific Purposes (ESP)   117

7 Value Addition to the Communication Course   119
   S. MEKALA AND C. HARISHREE

8 Poetry in the Engineering Curriculum   132
   S. P. DHANAVEL AND S. KUMARAN

9 Incorporating Thinking Skills in the Engineering Curriculum   144
   S. MEKALA AND C. HARISHREE

PART V
Approaches and Strategies   159

10 Pedagogical Strategies in Improving ESL Learners'
    Speaking Proficiency   161
    S. SHANTHA AND R. K. DHARINI

11 Teaching Speaking in Indian ESL Classrooms   189
    R. NANDHINI AND D. POORVADEVI

12 Improving Reading Proficiency of ESL Learners   201
    B. S. PRAMEELA PRIADERSINI AND C. HARISHREE

PART VI
Digital Education   223

13 Techno-Mediated Teaching of English for Engineers in
    India: Perceived Threats and Hidden Opportunities   225
    T. RAVICHANDRAN

14 Integration of Zone of Proximal Development (ZPD) and
    ICTs in Language Learning   237
    M. PONMANI AND GEETHA R

PART VII
Teacher Education   **253**

15 Towards Designing a Module for INSET   255
    K. MANJULA BASHINI

16 Reflective Practice for Teachers in the Post-Method Era   271
    S. SOUNDIRARAJ AND B. ANDRIA BABU

Appendix A   283
Appendix B   286
Index   290

Figures

2.1	P21 Framework of 21st Century Skills Set	22
2.2	Taylor's Hierarchy of Creativity	27
2.3	Tenets of Learning and Innovation Skills	28
3.1	Expectations of Employers vs ELP of Employees	51
6.1	Overall Performance Levels of all Participants in English Language Skills before the Study Period	111
6.2	Pre-Test Performance Percentage of all Participants in English Language	111
6.3	Overall Performance Levels of Participants in English Language Skills after the Study Period	112
6.4	Post-Test Performance Percentage of all Participants in English Language	113
11.1	Theory of Communicative Language	194
11.2	Illustration of Picture Description Task	196
11.3	CEFR Global Scale	199
14.1	Zone of Proximal Development	238
14.2	Gilly Salmon's Five-Stage Model	243

Tables

2.1 Traits Important to Critical Thinking 24
2.2 Communication Skills 25
2.3 Classification of Learning Strategies 29
2.4 Cognitive Strategy Descriptions and 4Cs 30
2.5 Metacognitive Strategy Descriptions and 4Cs 32
3.1 Employers' Views on Low ELP Affecting Reputation of the Organization 47
3.2 The Impact of Communication Skills on the Productivity of an Organization 48
3.3 Employers' Views on the Aspects of Communication That Lead to Productivity 49
3.4 Employers' Views on Skills Gap and the Reasons for It 50
3.5 Employers' Evaluation of Communication Skills during the Recruitment Process 51
3.6 Factors Leading to Skills Gap by Affecting Productivity 52
3.7 Employers' Opinion on the Remedial Measures 53
4.1 Definitions of Task 63
4.2 Three Approaches to TBLT 65
4.3 Difference between Traditional Methodology and Task-Based Methodology 66
4.4 Task-Based Lesson Plan: Ellis Model (2003) 67
4.5 Task-Based Lesson Plan Model 68
4.6 Taxonomy of Task Types in Willis and Willis (2007) 71
4.7 Main Differences Between Focusing on Language and Form 72
5.1 Descriptive Statistics of Pre-Test 92
5.2 Descriptive Statistics of Post-Test 92
5.3 One-Way ANOVA between Pre- and Post-test 92
6.1 Select Sub-Skills of LSRW in Integrated Format 101
6.2 Percentage and Performance Levels 110
10.1 CEFR Speaking Assessment Criteria 167
10.2 Pedagogical Strategies Employed in the Study 168
10.3 One-Way ANOVA for Pedagogical Strategy-I 171
10.4 One-Way ANOVA for Pedagogical Strategy-II 172

10.5	One-Way ANOVA for Pedagogical Strategy-III	173
10.6	One-Way ANOVA for Pedagogical Strategy-IV	174
10.7	One-Way ANOVA for Pedagogical Strategy-V	175
10.8	One-Way ANOVA for Pedagogical Strategy-VI	176
10.9	One-Way ANOVA for Pedagogical Strategy-VII	177
10.10	One-Way ANOVA for Pedagogical Strategy-VIII	178
10.11	One-Way ANOVA for Pedagogical Strategy-IX	179
10.12	One-Way ANOVA for Pedagogical Strategy-X	180
10.13	One-Way ANOVA for Pedagogical Strategy-XI	181
10.14	Correlation Analysis of Speaking and Learning Strategy Use	182
11.1	Illustration of Role-Play Task	197
12.1	Self-Assessment vs Actual Performance – An Analysis	213
12.2	Reading Speed and Comprehension: 1st Passage and Last Passage	214
12.3	Reading Speed and Comprehension – A Paradox	216
12.4	Retention Level of the Students	217
14.1	Sample Learning Tasks Based on Gilly Salmon's Five-Stage Model	246
14.2	Participation in e-Learning According to Level of Engagement	247
16.1	Experience-Wise Classification of Teachers	276

Preface

The primordial objective of engineering pedagogy is to capacitate the engineering graduates with the requisite knowledge and skills that would make them productive employees and competent individuals. The English language teaching strategies have always proven to be very efficacious in integrating such skills and simultaneously in augmenting the language skills. Consequently, the field of English Language Teaching (ELT) has undergone a paradigm shift in its methods and approaches of teaching and learning. In spite of the state-of-the-art developments, a gaping chasm (skills gap) still prevails between the competence of engineering students and the expectations of the industry. This book aims to explore the various possibilities of pedagogical practices in the engineering curriculum to bridge this gap, both theoretically and practically.

In this book, we have compiled the insights and ideas of various renowned and acclaimed professors and assiduous research scholars from diverse universities as well as prominent industry experts on different but interconnected themes revolving around ELT teaching methods and curriculum in engineering institutions. The chapters in this book comprise the contributors' research in a variety of Indian engineering education contexts in the English classroom. The 16 chapters in the book contributed by the 19 authors are organized into seven sections based on the following thematic categories:

- **Developing Competency and Skills** explores the need for remodeling curriculum that fosters higher levels of competency and skill in engineers so the learners are equipped and ready to face challenges in the workplace. It proposes the integration of learning strategies in the engineering classroom to assist professors in facilitating the learners' potential to cope with their workplace requirements.
- **Communication Skills** establishes the requirement of English communication skills and recommends strategies to bridge the skills gap and improve the performance and productivity of the individual and their organization. It also propounds task-based language teaching (TBLT) methodologies to improve the students' communication skills.

- **Listening, Speaking, Reading and Writing (LSRW) Skills** deals with the effectiveness of the Test, Teach, Test (TTT) method to improve learners' proficiency, especially of those from rural areas. It also posits conscious teaching of a set of sub-skills to create awareness in students to encourage them to move towards competency in their language and communication skills.
- **English for Specific Purposes** highlights the efficacy of Thinking Skill tasks and project-based learning in guiding the students to comprehend the nuances of communication. It advocates employing poetry in the context of English for Specific Purposes (ESP).
- **Approaches and Strategies** enunciates the influence of pedagogical strategies on the learning strategies and provides a practice-oriented approach in teaching Spoken English in English as a Second Language (ESL) context. It also scrutinizes the effectuation of reading strategies to improve the reading skill of ESL learners.
- **Digital Education** demonstrates how techno-mediated teaching facilitates reconfiguration of instructional design that focuses on the learner-centered mode and postulates the integrated practice of the Zone of Proximal Development (ZPD) and Information and Communication Technology (ICT) in the language learning process.
- **Teacher Education** asserts the significance of in-service education for the teachers and explicates the use of reflective practice with Gibbs' model to make the teachers autonomous in different aspects of second language teaching.

This collection is designed to highlight the imperative changes to be assimilated into the engineering curriculum for the learners to flourish in the global workforce. This book is primarily intended for English professors, teacher educators, policymakers, leading authorities, and research scholars in the Indian engineering education context. Nevertheless, we hope this book would be a valuable resource for educators and researchers of other disciplines worldwide.

Acknowledgments

We are honored and deeply grateful to Routledge Publications for providing us with this prodigious opportunity. We express our sincere thanks and great appreciation to Ms. Lubna Irfan, Associate Commissioning Editor, for her timely guidance, limitless patience, and continuous support throughout the project. Further, we are immensely grateful to all the reviewers, copyeditors, and staff, who have planted their diligent efforts to design, regulate, and productively present the book to the readers.

We owe an enormous debt of gratitude to all our contributors from IITs, NITs, State Universities and Colleges, whose incredible insights and invaluable work have immensely helped transform the vision of the book into reality.

We would like to express our extensive veneration and colossal gratitude to Ms. C. Harishree, for her assiduous work in strategically organizing and formatting the book chapters.

We would also like to extend our heartfelt thanks to Ms. T. Sangeetha, Ms. K. Priyadharshini, and Ms. K. Mohana Aishwarya for their immense help in the compilation of the book.

Contributors

B. Andria Babu is a Research Scholar in the Department of English at Anna University, Chennai, India.

N. Chitra is an Assistant Professor in the Department of English at University College of Engineering, Anna University, Tiruchirappalli, India.

S. P. Dhanavel is a Professor of English at Indian Institute of Technology Madras, Chennai, India.

R. K. Dharini is a Research Scholar in the Department of Humanities and Social Sciences at National Institute of Technology, Tiruchirappalli, India.

Geetha R is a Doctoral candidate who completed her PhD research focusing on the augmentation of learners' writing skill, in the Department of Humanities and Social Sciences at National Institute of Technology, Tiruchirappalli, India.

C. Harishree is a PhD researcher who completed her Doctoral degree specializing on 21st Century Skills in the Department of Humanities and Social Sciences at National Institute of Technology, Tiruchirappalli, India.

P. Hiltrud Dave Eve is an Assistant Professor in the Department of English at As-Salam College of Engineering and Technology, Thirumangalakudi, India.

S. Kumaran is an Assistant Professor in the PG & Research Department of English at Thiruvalluvar Government Arts College, Rasipuram, India.

K. Manjula Bashini is an Associate professor in the Department of English at Kumaraguru College of Liberal Arts and Science, Coimbatore, India.

S. Mekala is an Associate Professor and Head in the Department of Humanities and Social Sciences at National Institute of Technology, Tiruchirappalli, India.

R. Nandhini is an Assistant Professor in the Department of English at BIT Campus, Anna University of Technology, Tiruchirappalli, India.

B. S. Prameela Priadersini is an Assistant Professor in the Department of English at Government Arts College for Women, Pudukkottai, India.

M. Ponmani is an Assistant Professor and Head in the Department of English at Thassim Beevi Abdul Kader College for Women, Kilakarai, India.

D. Poorvadevi is a Director-Academics at Ebek Language Laboratories Private Limited, Chennai, India.

T. Ravichandran is a Professor of Department of Humanities and Social Sciences at Indian Institute of Technology, Kanpur, India.

C. Shanmuga Priya is an Assistant Professor in the Department of English at SRM Institute of Science and Technology (Deemed to be University), Tiruchirappalli, India.

S. Shantha is an enterprising woman of letters who has completed her PhD research with a focus on improving the speaking proficiency of engineering students, in the Department of Humanities and Social Sciences at National Institute of Technology, Tiruchirappalli, India.

S. Soundiraraj is a Professor in the Department of English at College of Engineering, Guindy, India.

R. Vasanthan is an Assistant Professor in the Department of English at National College, Tiruchirappalli, India.

Part I

Developing Competency and Skills

Chapter 1

Need for Remodelling the Engineering Curriculum

C. Harishree

The expectations and exigencies at the workplace have seen a momentous transformation from late 20th Century to early 21st Century. With the development of digital technologies in the 21st Century, there has been a transformation in business and in the workplace environment. The repercussions of the COVID-19 pandemic have vehemently increased the necessity of digital technologies in peoples' lives. To survive in this global challenge and meet the unwitnessed job requirements, students have to be prepared with essential digital skills. 21st Century skills are the set of abilities encompassing learning skills, literacy skills, and life skills. They are the most requisite skills set for an individual to thrive in the information age. In the education field, 21st Century skills are broadly used to refer to the set of skills or abilities obligatory for the students to be work-ready individuals. Indian education has been insisting to develop the skills set of the students and teachers to meet the desideratum of the world of work. The digital transformation of the present global workplace demands its workforce to be equipped with the additional skills set along with the subject knowledge. Kampen (2019) has defined 21st Century skills as a "range of competencies taught across all levels of education, that give students the skills they need to navigate an ever-shifting workforce". Likewise, the International Association for the Evaluation of Educational Achievement (IEA) is presently conducting a study on 21st Century skills entitled as "21CS MAP". It aims to define and implement 21st Century skills in the education field by comparing the curriculum of different countries and analysing the skills incorporated in each curriculum. It has invited the international community of education to participate in the study. The timeline of the IEA study has been proposed between February 2020 and December 2021. Similarly, 21st Century skills has been analysed around the world to prepare the students for the competitive edge of the future workplace situation. In this line, the chapter will discuss the different frameworks of 21st Century learning and the significance of Partnership for 21st Century Skills (P21) framework in the current scenario of Indian engineering education.

DOI: 10.4324/9781003268529-2

The students of engineering in India are facing a massive employment degeneration due to skills shortage. According to the annual employability survey of Aspiring Minds, "80% of engineers are not employable for any job in the knowledge economy" (Aggarwal, Nithyanand & Sharma, 2019, p. 5). The report further states, "The students should receive academic counselling to make them aware of their skills gaps and to provide them with a support strategy for addressing those gaps" (Aggarwal et al., 2019, p. 7). Accordingly, the skills mismatch between the industrial requirement and students' skill set has been studied by disparate researchers, and variegated list of skills has been identified as requisite skills set for the students of engineering. Bano and Vasantha (2019) have stated that skills gap arises because of the digitalization and invention of new technologies in the global workplace. The curriculum is not updated at a requisite pace according to the need of these digitalization and technological development of the workplace. Ravichandran and Abirami (2017) have studied the industry-expected skills and students' skills set through mailing system and direct contacts using questionnaire. The authors have also collected secondary data through published works and have explored few skills set necessary for employment. Despite these studies, the skills gap still persists due to the non-holistic approach on the essential skills set for the workplace preparedness of the students. The skill development training is also not sufficient for the students' career readiness. It may be due to the lacuna in the engineering curriculum being focused more on improving the technical skills of the students. Hence, the chapter accentuates the need for integrating 21st Century skills in the engineering curriculum and recommends to incorporate this skills set in the English classroom of the students of engineering to have more impact on fostering the students' skills set. In addition, the chapter explores the methods and approaches that could help in integrating the 21st Century skills in the engineering classroom. Concisely, the chapter expounds the significance of 21st Century skills in the workforce readiness of the students of engineering.

Frameworks of 21st Century Skills

There are diversified definitions for 21st Century skills, and different frameworks have been proposed by various organization to be incorporated in the curriculum. The common goal of all the frameworks is to encompass basic workplace skills such as critical thinking, problem solving, creativity, communication, and teamwork. UNESCO report by Delors et al. (1996) comprises four pillars of education named as 'Learning to know', 'Learning to do', 'Learning to be', and 'Learning to live together'. It is an important backdrop of lifelong learning. It has prompted the identification of the contemporary prerequisites of education for the changing needs of the workplace. Subsequently, different frameworks have emerged on skills set constituting towards the tenets of 21st Century skills. Dede (2009) has compared P21 framework (2009) with other frameworks namely enGauge framework of Metiri Groups and NCREL (2003), the American Association of colleges and Universities (AACU) framework (2007), the International Society for Technology in Education (ISTE) (2007) ICT Skills, and Assessment and Teaching of 21st Century Skills (ATC21s) (2010). It has been evident from Dede's comparison that P21 framework will be more effective in the career readiness of 21st Century students, as it encompasses the all-inclusive skills set of other frameworks. Additionally, the European Commission (2007) framework, the World Economic Forum (2015) framework, and the European Framework for the Digital Competence of Educators (DigCompEdu) (Redecker, 2017) are the most significant frameworks of 21st-Century learning for the workplace preparedness of the students of the 21st Century. Furthermore, Wagner (2014) has proposed seven surviving skills for facilitating the 21st-Century students towards career readiness. In addition, the Australian Council for Education Research (ACER) has published a framework (McLean et al., 2012) with five core skills and individual skill development frameworks for critical thinking, creativity, and collaboration skills. Furthermore, there are few frameworks concentrating on set of skills required for the specific circumstances. On the contrary, the P21 framework encompasses the holistic approach towards the requisite skills set to prepare the students for the world of work. Anagun (2018) has emphasized "the framework created by Partnership for 21st Century skills (P21, 2009) has been more widely adopted because of its thematic relevancy for the 21st Century and well-structured design" (p. 826). In this regard, it is evident that the P21 framework is deftly adaptable to the engineering field and can provide necessary skills set for the workplace readiness of the students of engineering.

P21 Framework

The partnership for 21st-Century learning is an organization formed in coalition with experts, business leaders, corporates, academicians, and policy makers. The US Department of Education, Microsoft, Cisco, and the National Educational Academy of US are part of the P21 organization. In 2018, the P21 organization has merged with Battelle for Kids. Since then, the Battelle for Kids organization has been trying to integrate the P21 framework with the academic content of school education. The P21 framework has been used by more than 220 schools in the US and the countries around the world and by the educators and policy makers, education departments, and so on. The purpose of the framework is to enhance the students' skills set to compete in the constant changing world of work. The framework has emphasized on teaching the recommended 21st Century skills along with the core subjects. The P21 framework can be easily adapted due to its comprehensive view of career ready skills. The P21 organization has stated that the framework will hone the students' engagement in the learning process eventually ensuring them to graduate better prepared for thriving in today's global economy (P21, 2009). Apart from the core subjects, the P21 framework categorized 21st Century skills into learning and innovation skills, information media and technology skills, and life and career skills.

Learning and Innovation Skills

Learning and innovation skills are considered quintessential irrespective of the field of study due to its attributes and inevitability in the workplace. It consists of creativity and innovation, critical thinking and problem solving, communication and collaboration which are commonly called four C's. The four C's are critically examined and implemented in the frameworks of school and college curriculum around the globe for its effectiveness in preparing the students to have competitive edge in the world of work. Trilling and Fadel (2009) have defined four C's as

> the ability to ask and answer important questions, to critically review what others say about a subject, to pose and solve problems, to communicate and work with others in learning and to create new knowledge and innovations that help build a better world.
>
> (p. 49 & 50)

It is vital for the students of engineering to imbibe this skills set to thrive in their workplace environment. Accordingly, to develop the four C's of the students of engineering, there is a need for a paradigm shift in the traditional content-based teaching by incorporating the four C's – creativity, critical thinking, collaboration, and communication – that ultimately will equip the students of engineering to identify innovative solutions to the real-world problems, investigate the projects, and collaborate with others to achieve the challenging tasks.

Creativity and Innovation

The P21 framework has defined the characteristics of creativity and innovation, as employing idea creation techniques like brainstorming, creating incremental and radical concepts, being receptive towards diverse perspectives, exhibiting inventiveness, contributing to the field of work by working on novel ideas, and implementing it at workplace. Creativity is not limited to particular field of study. In the 21st Century, demand for creativity has been steadily increasing as the digital technologies has abundant information and readymade solutions for critical situations. Hence, organizations expect its workforce to produce novel ideas and solutions for their products and in crucial situations at workplace.

Critical Thinking and Problem Solving

Critical thinking skills can be effective in motivating the learning outcomes of the students. The necessary thinking process for applying one's knowledge at particular situation needs critical thinking skills. Thenceforth, P21 defines attributes of critical thinking and problem-solving skills as effective reasoning, decision making and judgements, conventional and innovative ways of problem solving, and clarifying different aspects of a problem with significant questions. Higher order thinking skills of Bloom's taxonomy are also categorized under the critical thinking skills of the P21 framework. Critical thinking will reflect on new insights of the information and attitude towards lifelong learning. In the same way, problem-solving skills are emphasized as the world of work that meet unpredictable problems in the changing global scenario. For instance, the COVID-19 pandemic is a best example as the workplace has shifted from physical environment to digital platforms. Industry needs its workforce with more emphatic skills to solve the novel critical situations in the workplace.

Communication

In the knowledge-based society of the 21st Century, communication plays a crucial role in exhibiting the competence and thoughts. Communication is required in both verbal and non-verbal mode of delivery. In the present age, the non-verbal communication shares equal importance with verbal communication in different forms and contexts. Listening, interpreting, inferring, expressing, and implying the content or information are some of the key features of communication skills in the P21 framework. The framework has emphasized that communication creates an impact on the students' content knowledge. Communication parameters that can be implemented and measured in the language classroom have been briefly described in the Common European Framework of Reference for Languages (CEFR) (Council of Europe, 2001). Congruently, implementing communication skills in the 21st-Century classroom will help the students to present their fluency of content and their ability to perform better according to the needs of the workplace environment.

Collaboration

Digital technologies have integrated the workplaces globally into a single platform eventuating the students of 21st Century to work with different group of people and produce products satisfying global customer needs. Hence, the collaboration skills have become essential among the students. Collaboration skills denotes the ability to work effectively with diverse group of people. Team management, shared responsibility, and individual contribution to the team are the key aspects of collaboration skills. Online platforms are the best examples of collaborative workplace comprising the information uploaded by people from different countries and from diverse socio-cultural background. Hence, upgrading the students' collaboration skills will enable them to understand the different point of views of their teammates across the globe and increase their productivity at their future workplace.

Information, Media, and Technology Skills

Information, media, and technology skills are also called digital literacies. Information literacy is defined as the ability to "locate, evaluate and use effectively the needed information" (American Library Association, 1989). The P21 framework categorizes information literacy, media literacy, and ICT literacy (information, communications, and technology) under information, media, and technology skills. It is broadly known as literacy skills. In the modern world of digital technology, it becomes purposeful for the students of 21st Century to handle the technology efficiently and obtain the requisite information pertaining to their field of study. The rapid changes in the world have increased the necessity for employees to contribute to their organization with effective use of technology. Congruently, digital technologies have made fundamental changes in the interaction and working module of the 21st Century workplace. So, the students of engineering need to explore the copious content available in the digital platforms and have to identify the authentic and related information for the circumstance of their place of work. Eventually, this skills set will help the students of engineering to develop their future career prospects.

Information Literacy

Information literacy refers to accessing, critically evaluating, and applying the researched information creatively for solving issues or problems at real-life situations. It requires "fundamental understanding of the ethical/legal issues surrounding the access and use of information" (P21, 2009). In the internet-based society, book knowledge will not suffice to possess competitive edge in the global economy. The information literacy thus helps the students acclimatizing the demands of information era and enhancing their skills to explore the essential information.

Media Literacy

The P21 framework has defined the characteristics of media literacy as the cognizance of message conceptualization in media besides interpreting its purpose, tools, features, and conventions. In addition, examining the expositions of content by set of people from different socio-cultural background and its impact on beliefs and behaviours of the people symbolizes media literacy. Further, media plays a crucial role in the perspicacity of the product information influencing the perspectives of the consumers. Media has become a part of a personal and professional life of every individual. Hence, 21st-Century workforce expects its employers to be able to comprehend and use media to improve the productivity of the organization and upsurge the product reach to the customers.

ICT Literacy

The 21st Century is also called the digital age due to the prevalent usage of computers and digital gadgets in work and life. There are very few exceptional workplace environments, where digital gadgets do not play a major role. Children born in the 21st Century wielding advanced digital tools from their childhood unlike older generations are called 'digital natives'. As a consequence, in order to thrive in this 21st Century, students need to be more techno-savvy at workplace. Correspondingly, in ICT literacy, the P21 framework engenders the application of technology effectively for research, communication, organization of information, and function effectively in our knowledge economy.

Life and Career Skills

Life skills have been emphasized to be a part of the engineering curriculum from the end of 20th Century by educators, policy makers, researchers, and experts in congruence with the asseveration by the National Curriculum Framework (2005) of India to integrate life skills in the curriculum for the better career prospects of the students. Life skills have various traits according to the necessity of the workplace situation and the field of study. Life and career skills of the P21 framework, the inevitable tenets for personal and professional success, is the "Combination of career & learning self-reliance competence" for the overall productivity of peers (Lichtenegger, 2014). It consists of five skills set namely flexibility and adaptability, initiative and self-direction, social and cross-cultural skills, productivity and accountability, and leadership and responsibility. Hence, developing life and career skills will promote the students of engineering to meet the complex demands of the global workplace and to succeed in their professional and personal life by bridging the skills gap in the engineering field and proliferating the employment opportunities of the students of engineering.

Flexibility and Adaptability

According to the P21 framework, flexibility can effectuate the students to negotiate different views and believes, to infer criticism positively, and to incorporate the feedback for productive work results. Similarly, these skills help the students in adapting to different roles and responsibilities in the workplace and in producing quality work outcomes even at ambiguous situations. Employees with flexibility and adaptability skills will be potentially adequate to work with different teams and at different working areas benefitting the organization to compete in the global economy. Therefore, students capacitated with flexibility and adaptability skills are competent to work in all situations and to manage different roles in the workplace.

Initiative and Self-Direction

Goal management, time management, self-initiation, and self-engagement towards the learning process are some of the key attributes of initiative and self-direction skills of the P21 framework. The framework also suggests to promote skills set and achieve expertise beyond curriculum. This skills set can help the students to initiate learning and develop their knowledge autonomously. The basic knowledge of the students obtained from their course of study need to be expanded further in their workplace to achieve high positions in their profession necessitating initiative and self-direction skills for all the students of the 21st Century to self-regulate their knowledge and act autonomously.

Social and Cross-Cultural Skills

Social and cross-cultural skill sets are related to collaboration skills enabling the students to work and interact effectively in a team by understanding the socio-cultural background of their team members. Further, social and cross-cultural skills equip the students to bridge cultural differences within a group, and be receptive to values and ideas of teammates for innovative and quality work outcomes. As discussed earlier, COVID-19 has transformed the physical workplace environment into digital workplace and people from all over the country are made to work collaboratively in a digitized platform. People with various cultural values and social beliefs are entailed to work with each other for creative and innovative solutions and development. Hence, developing the social and cross-cultural skills will enhance the opportunities for the students in global work market.

Productivity and Accountability

Productivity is directly connected with project management. The P21 framework propounds goal setting, achieving the standard goals, and prioritizing the work to deliver the intended result for the students of the 21st Century. Further, P21 (2009) enlists the tenets of productivity and accountability skills as "be accountable for results", "reliable and punctual", "respect and appreciate team diversity", and "work positively and ethically". Companies and firms are dependent on its human resource for growth and success. Therefore, the 21st Century demands a workforce capable of immense productivity and accountability in their entrusted work. Thus, empowering students with productivity and accountability skills will promote their contribution to the prospective organizations, achieving great success in their career.

Leadership and Responsibility

It is important to learn responsibility at earlier stage of career to step ahead in the job. The communication of leadership and responsibility is essential, as it is necessary to lead or guide others with responsibility. The P21 framework has defined the leadership and responsibility skills as guiding and leveraging people towards a common goal, and maintaining integrity among the team members. Further, responsibility is characterized with "larger community in mind" (P21, 2009). It is desideratum to accentuate the students of the 21st Century with the leadership and responsibility skills to lead a group and to act with responsibility at critical situations in the workplace.

Engineering Education

Engineering graduates in India has been facing steady decline in their employment opportunities in recent years. Engineering education in India was established during pre-independence era with civil engineering as the primary engineering field. At present, there are different disciplines of engineering and the prerequisites of each discipline have also evolved over the years. Subject knowledge and technical skills have become inadequate for the career opportunities in the present information age. The students are expected to possess the ability to solve critical problems and collaborate with diverse group of people in addition to work at multicultural and multinational circumstances. It is a seemingly daunting task for engineering institutes to provide holistic knowledge required for the students' profession as the skills taught are often outdated due to rapid changes in the industry. This lag has led to the skills deficit among the students of engineering for the 21st-Century workplace. In this line, the All India Council for Technical Education (AICTE) has been established in 1988 to monitor the technical education in the country and update the teaching-learning process to suit the industrial demands. The AICTE has framed different committees to analyse the development of technical education along with the growing need of global economy. These committees have analysed the data of the AICTE "on engineering capacity and enrolment trends across the country" (Reddy, 2019, p. 2). The committees' report has identified the gap areas and issues in the engineering education in India, has recognized the prevailing skills gap, and has promulgated recommendations to overcome the issues. Likewise, the survey reports of Aspiring Minds (2016) and National Skill Development Corporation (NSDC) (2007) have examined the reasons for lack of employment of the students of engineering and have found that the students do not possess skills set expected by the employers to thrive in the global competitive workplace. These reports implicitly emphasize on skills mismatch between the industry and academia.

Similar studies have examined the employability of the students of engineering in India from employers' perspective. For instance, Srividhya and Vijayakumari (2017) have stated "India is one of the largest producers of engineers in the world". The authors have further stated the opinion of CoreEL Technologies that only 15%–20% are employable among 6 lakhs of the engineers and have professed the statement of Aspiring Minds that only 7% in 1.5 lakhs of the engineering students are fit for employment in India. Furthermore, the employers' perception on the requisite skills set has been studied by researchers and policy makers. Ravichandran and Abirami (2017) have examined the gap between the expected skills set of the workplace and the skills possessed by the students of engineering.

It becomes the responsibility of the institutions' curriculum designers and policy makers to bridge this skills gap. Moreover, Unni (2016) has affirmed "the current education system does not adequately focus on skills which can improve employability, and large section of labour possess outdated skills". In this regard, there are organizations like PeopleStrong, Wheebox, CII, etc., to examine the requisite workforce skills set to be incorporated in the higher education. These organizations further investigate the lacunae in the education system to improve the quality of the education and prepare the students to progress towards the global economy. It is evident from the above studies that there is a profound skills gap between the demands of the industry and the students' skills set in the engineering field. In this regard, organizations like the AICTE and NSDC concentrate on enhancing the teachers' qualification and skills set through various programmes to meet the educational challenges of present digital era. But there is no effective measure to integrate 21st Century skills in the engineering curriculum for fostering the students' skills set and preparing them for their better career prospects.

Need for Integrating 21st Century Skills

Organizations have expressed their dissatisfaction towards students' career readiness; whereas, business leaders around the world have stated that they are not able to find employees with requisite skills set to support their customers' needs. Many organizations also train the new recruits to fill the skills gap in the industry. Ramanan, Kumar, and Ramanakumar (2015) have acclaimed that 80% of engineering employability is in high defect rate in India, as per the literature survey and "under the globalized scenarios of industrial production, large employment opportunities are awaiting fresh engineering graduates, if they meet industry expectations". Warner and Kaur (2017) have also opined "facilitating learning at all levels of the education stratum to create effective 21st century knowledge creators, inventors and innovative workers is increasingly recognized today as a primary objective of education" (p. 193).

The goal of education will not suffice by providing literacy to the students of the 21st Century who are in an indispensable need to equip themselves with requisite skill set and competencies for employability. Skills development has gained advantage over the years to sustain competitiveness among the organizations in the global economy. The corporate sectors of the 21st Century requires highly skilled workers to meet the global competition necessitating the students of engineering to hone their skills set and satiate the towering demands of the world of work. The nature of work has been evolved with technology advancement that entails the workforce to be equipped with the advanced skills set. Maclean and Ordonez have remarked that employers seek for more 'trainable recruits' for their organization (as cited in Raftopoulos, Coetzee, & Visser, 2009). The curriculum frameworks also insist on skill development. Yet there are no proper guidelines for incorporating the requisite skills set in the engineering curriculum. Consequently, researches have been undertaken identifying the important skills to be acquired by the students for the 21st-Century workplace and proposing the integration of those skills in the curriculum framework of the students.

Countries like America, Australia, Austria, Finland, etc., have taken measures in preparing the students to meet the requisiteness of present competitive world of work. The students are being trained on 21st Century skills along with their subject knowledge in these countries. Schools in Madeira and Virginia have included the 21st Century skills of the P21 framework in their curriculum and accentuated the efficacy of the P21 framework in enhancing the students' skills set. Microsoft Partners in Learning and the Pearson Foundation and Gallup (2013) have explored the relationship between 21st Century skills in the classroom and quality of work at workplace. They have proposed the P21 framework to be blended with classroom teaching to improve the students' skills. Similarly, the Central Board of Secondary Education, Delhi (2020), has published a handbook on 21st Century skills with proposed activities for implementing in the classroom of Indian schools and develop the students' skills set. Hence, it is evident that the need for 21st Century skills is vital in the present education system.

Apparently, the skills gap reports in India have emphasized on promoting skills development programmes and professional development courses to increase the probability of obtaining employment opportunities. Yet, there is no report to define the comprehensive skills set for the workplace preparedness of the students of engineering in India. The 21st Century skills set have been considered, while deliberating the constituents of work-related skills like soft skills, life skills, employability skills, professional skills, transferrable skills, etc. It has been metamorphosed into future skills with reference to Industry 4.0 technologies and necessities. Yet, there is no intact list of skills applicable to all engineering fields for providing comprehensive approach to skills development of the students of engineering. Further, the paucity in the engineering curriculum hinders the effectual remedies taken by the government and the educational institutions towards the holistic workplace preparedness of the students. Scanty studies have been conducted on implementing 21st Century skills in the engineering classroom. So, it is necessary to integrate the 21st Century skills of the P21 framework in the engineering curriculum, which can provide an aggregated overview of the necessary skills and augment students skills set for their greater career possibility. Correspondingly, integrating 21st Century skills into the engineering curriculum will also enable the students to develop their skills set along with their subject knowledge. Henceforth, the engineering education in India need to improvise its core curriculum by integrating 21st Century skills, as it will enable the student beneficiaries to succeed in their personal and professional life endowing them 'economically productive population' in the society.

Integrating 21st Century Skills in the English Classroom

In recent years, the language classrooms have transformed its attention towards the employability of the students. Most of the engineering institutes in India have employed the English teachers to develop the students' language skills according to the need of the industrial requirement. For example, communication skills have received major deficit in the place of work, as the students of engineering are facing big trouble in communicating their ideas and thoughts in English language at workplace environment. The skills gap in communication has been broadly researched and engineering institutes have introduced specific English courses emphasized on professional communication. Communication skills is one of the skill sets among other skill sets that the students of engineering are found inadequate bespeaking the need to hone requisite workplace skills. Comprehensively, the workplace environment requires the students to be more flexible, reliable, productive, creative, collaborative, etc. Hence, the 21st Century skills encompassing all the attributes of workplace environment should be integrated with the English language syllabi that can promote the students' skills set along with the language skills. Moreover, LSRW skills of the students of engineering has been developed by various skill-based activities in the English classroom of the engineering institutes like, group discussion, role plays, debates, and seminar presentations. In addition, Lichtenegger (2014) has studied the P21 framework in the EFL context of Austria and has propounded approaches and methods to equip the students with 21st Century skills in the EFL classroom. In this line, Lichtenegger has explicated methods to implement 21st Century in the English classroom for effective workplace preparedness of the students that can be adapted to equip the students of engineering with 21st Century skills. Moreover, constituents of P21 framework are also associated with language skills. Hence, teachers facilitating the students' English language skills can effectively integrate the 21st Century skills into their pedagogy and constructively facilitate career ready students of engineering.

Methods and Approaches to Integrate 21st Century Skills

To integrate the skills in the curriculum, the content proposed to be taught needs evaluation from the perspectives of students and teachers. Anagun (2018) has examined the perceptions of primary school teachers' proficiency in the 21st Century skills and constructivist learning in the province of Eskisehir, Holland. The author has acclaimed that "teachers are key actors who shape students' learning and have a critical role in implementing new approaches to learning". It evinces the teachers' beliefs on the new instructional methods and curricular designing to be essential. Warner and Kaur (2017) have proposed 2T2C (Thinking, Technology, Communication. and Confidence) model perceiving students' and teachers' opinion on its effectiveness for mathematics instructional model. Their work is a part of curriculum design in the Republic of Trinidad and Tobago (West Indies) where this model has been implemented in the secondary school curriculum comprising all the industry requirements and cognitive levels in the Bloom's taxonomy in addition to teacher training programme. Yet, the model couldn't achieve its goal at an expected level due to the conventional teacher-centred instruction method. Hence, Warner and Kaur have suggested learner-centred instruction for the successful 2T2C model. Fatmawati (2018) has proposed project-based learning by conducting survey among the students of English department of Private University in East Java. Further, the author has suggested educators to find 'whether students possess skills for the 21st Century such as problem solving, critical thinking, creativity and entrepreneurship' and assign 'project-based learning with specific goals of 21st Century skills development'.

Ganayem and Zidan (2018) have evaluated the students' perceptions on the role of online instruction in developing 21st Century skills. They have also examined the students' preference on learning style like 'face to face, synchronous and asynchronous' and have found that the majority of the students preferred face to face learning style performing well on ICT skills, besides improving their communication and collaboration skills. Similarly, collaborative approach, project-based approach, and problem-based approach are some of the approaches that can be implemented in the classroom to develop the students' skills set. Sanabria and Aramburo-Lizarraga (2017) have proposed Gradual Immersion Method (GIM) to prepare the students for the information- and knowledge-based digital age. The GIM is based on creative learning with the use of augmented reality and digital interactive devices. Likewise, role plays, personal development activities, time-management tasks, recap activities, group dynamic tasks, problem-solving scenarios, and creative content tasks can be implemented in the

English classroom to enhance the students' skills set. The students of engineering can also be assigned to work with technologies to understand complex aspects and to promote novel solutions with their group or team. It will induce the technology-integrated learning and can promote the ICT skills along with four C's. Tuzlukova, Busaidi, Burns, and Bugon (2018) have explored the perceptions of the 21st Century skills among the teachers in their teaching-learning process and have suggested the notional-functional approach and learning strategies instruction as effective approaches to teach critical thinking skills in the English language classroom.

Integration of the 21st Century skills in the classroom would propel the teachers of English to face certain constrains like completion of the syllabus on time, difficulty in obtaining students' involvement in the tasks and activities, assessment criterion for the examination, etc. Time constraint might be the prominent issue that the teachers have to complete the syllabus in the provided teaching timeframe. Therefore, while incorporating the 21st Century skills in the curriculum, the constraints of teachers to integrate the skills in their classroom teaching has to be evaluated by deliberating the suggestions of the teachers. Moreover, to overcome these constraints, the curriculum can incorporate 21st Century skills as an individual course and can be rendered in the semester prior to the recruitment process for the students of engineering. It can facilitate the teachers with sufficient time to nurture the students' skills set and students can functionally be assessed based on the tenets of the P21 framework.

Overview

This chapter has examined the 21st Century skills and its framework to give an overview of the concept and its impact on the engineering education. Further analysis of the P21 framework has explicated the constituents of the framework and its effectiveness in preparing the students for the 21st-Century world of work. As number of studies and reports have highlighted the skills gap of the students of engineering, it is conceptualized that engineering curriculum needs a revamping structure to meet the industrial requirements by incorporating 21st Century skills. In this line, the need for 21st Century skills for the students of engineering has been acknowledged with certainty in this chapter and consequently, the need for integrating 21st Century skills in the engineering curriculum has been explored to facilitate the students' skills set towards the demands of globalized workplace. Besides, it has also emphasized the implementation of the 21st Century skills in the English classroom to improve the efficacy of the integrated practice in the engineering curriculum. In order to make it congenial for the teachers to integrate the skills set, the chapter has reviewed the methods and approaches proposed by researchers. Further, it has suggested to incorporate the 21st Century skills as an individual course to annihilate the constraints that the teachers of English would face during the integration of the skills set in their pedagogical practices. Concisely, the chapter highlights the overview of frameworks, need, and implementation of 21st Century skills in the engineering curriculum for preparing the students to meet the challenges of their future workplace.

References

Aggarwal, V., Nithyanand, S., & Sharma, M. (2019). National employability report – Engineers, annual report-2019. Aspiring Minds. Retrieved from https://www.aspiringminds.com/research-reports/national-employability-report-for-engineers-2019/

American Library Association. (1989). Presidential committee on information literacy: Final report. Retrieved from http://www.ala.org/acrl/publications/whitepapers/presidential

Anagun, S. S. (2018). Teachers' perceptions about the relationship between 21st century skills and managing constructivist learning environments. *International Journal of Instruction, 11*(4), 825–840.

Aspiring Minds. (2016). National employability report: Engineers. Retrieved from https://www.aspiringminds.com/research-reports/national-employability-report-for-engineers-2016/

Bano, Y., & Vasantha, S. (2019). Review on strategies for bridging the employability skill gap in higher education. *International Journal of Recent Technology and Engineering (IJRTE), 7*(6S5), 1147–1152.

Central Board of Secondary Education, Delhi. (2020). *21st Century Skills: A Handbook.* Retrieved from http://cbseacademic.nic.in/web_material/Manuals/21st_Century_Skill_Handbook.pdf

Council of Europe. (2001). *Common European framework of reference for languages: Learning, teaching, assessment.* Cambridge: Cambridge University Press.

Dede, C. (2009). Comparing frameworks for "21st century skills". In J. Bellanca & R. Brandt (Eds.) *21st century skills: Rethinking how students learn*, pp. 51–76. Bloomington, IN: Solution Tree Press.

Delors, J. et al. (1996). *Learning: The treasure within.* France: UNESCO. Retrieved from https://docs.google.com/viewer?a=v&pid=sites&srcid=ZGVmYXVsdGRvbWFpbnxiYnNtZWRpYXBsYW5uW5nfGd4OjNlZGGJiM2Q0NTEzYmY2YmM

European Commission. (2007). *Key competencies for lifelong learning: European reference framework.* Luxembourg: Office for Official Publications of the European Communities.

Fatmawati. A. (2018). Students' perception of 21st century skills development through the implementation of project-based learning. *Pedagogy: Journal of English Language Teaching, 6*(1), 37–46.

Ganayem, A., & Zidan, W. (2018). 21st century skills: Students perception of online instructor role. *Interdisciplinary Journal of E-Skills and Lifelong Learning, 14*, 117–141. DOI: 10.28945/4090

Kampen, M. (2019). *How to promote 21st century skills in your school.* India: Prodigy. Retrieved from https://www.prodigygame.com/main-en/blog/21st-century-skills

Lichtenegger, B. (2014). 21st century skills-status quo in Austrian HLW EFL classrooms and implications for teaching (M.Phil. Dissertation). Retrieved from http://othes.univie.ac.at/33173/

McLean, P. Perkins, K., Tout, D., Brewer, K., & Wyse, L. (2012). *Australian core skills framework: 5 core skills, 5 levels of performance, 3 domains of communication*. Australia: Commonwealth of Australia. Retrieved from https://research.acer.edu.au/cgi/viewcontent.cgi?article=1011&context=transitions_misc

Microsoft Partners in Learning and the Pearson Foundation. (2013). *21st century skills and the workplace: A 2013 Microsoft partners in learning and Pearson foundation study*. Washington, DC: Gallup.

National Curriculum Framework. (2005). India: National Council of Educational Research and Training. Retrieved from https://ncert.nic.in/pdf/nc-framework/nf2005-english.pdf

National Skills Development Corporation. (2007). *Human resource and skills requirements in the IT and ITeS sector (2013–17, 2017–22)*. KPMG Advisory Services Pvt. Ltd. Retrieved from https://nsdcindia.org/sites/default/files/IT-and-ITeS.pdf

P21. (2009). P21 framework definitions. *Partnership for 21st Century Skills*. Retrieved from https://files.eric.ed.gov/fulltext/ED519462.pdf

Raftopoulos, M., Coetzee, S., & Visser, D. (2009). Work-readiness skills in the Fasset sector. *SA Journal of Human Resource Management, 7*(1), 119–126. DOI: 10.4102/sajhrm.v7i1.196

Ramanan, L., Kumar, M., & Ramanakumar, K.P.V. (2015). Approach towards reducing soft skill gap of engineering graduates in India from employers' perspective to employability. *International Journal of Engineering Research Online, 3*(6), 504–512.

Ravichandran, M., & Abirami, P.G. (2017). A gap between employers expectations and engineering students level in employability skills. *IJARIIE, 3*(3), 2280–2298.

Mohan Reddy, B. V. R. (2019). *Engineering education in India- short- & medium-term perspectives*. AICTE. Retrieved from https://www.aicte-india.org/sites/default/files/Short%20Term%20and%20Medium%20Term%20Report.pdf

Redecker, C. (2017). European framework for the digital competence of educators: DigCompEdu. In Punie, Y. (Ed.) *EUR 28775 EN*. Lusembourg: Publications Office of the European Union. doi: 10.2760/159770.

Sanabria, J. C., & Aramburo-Lizarraga, J. (2017). Enhancing 21st century skills with AR: The gradual immersion method to develop collaborative creativity. *EURASIA Journal of Mathematics Science and Technology Education, 13*(2), 487–501.

Srividhya, M. A., & Vijayakumari, D. G. (2017). Employability trends of engineering graduates in Tamil Nadu. *Economics, 3*(6), 65–66.

Trilling, B., & Fadel, C. (2009). *21st century skills: Learning for life in our times*. San Francisco: Jossey-Bass.

Tuzlukova, V., Busaidi, S., Al Burns, S., & Bugon, G. (2018). Exploring teachers' perceptions of 21st century skills in teaching and learning in English language classrooms in Oman's higher education institutions. *The Journal of Teaching English for Specific and Academic Purposes, 6*, 191–203. doi: 10.22190/JTESAP1801191T.

Unni, J. (2016). Skill gaps and employability: Higher education in India. *Journal of Development Policy and Practice, 1*(1), 1–17. doi: 10.1177/0000000315612310

Wagner, T. (2014). *The global achievement gap: Why even our best schools don't teach the new survival skills our children need – and what we can do about it.* Updated Edition. New York: Basic Books.

Warner, S., & Kaur, A. (2017). The perceptions of teachers and students on a 21st century mathematics instructional model. *International Electronic Journal of Mathematics Education, 12*(2), 193–215.

World Economic Forum. (2015). *New vision for education: Unlocking the potential of technology.* Retrieved from http://www3.weforum.org/docs/WEFUSA_NewVisionforEducation_Report2015.pdf

Chapter 2

Integrated Practice of Learning Strategies in the Engineering Curriculum

Geetha R

Skills are imperative requisites determining the prospective career of the engineering graduates. "Skills are the building block of human capital and the necessary catalyst required for development and economic prosperity of a nation" (Gaurav, 2020). It regulates students' cognition to be prepared for their challenging workplace situations. According to Agarwal (2006), India has the third largest higher education system in the world next to China and the United States in terms of the number of institutions. However, higher education institutes are falling behind when it comes to facilitating and fostering the requisite skills set for the engineering students that promote them to survive the unanticipated challenges in a workplace environment. This negligence has resulted in the inadequate students' competence to grapple with the job market and seek employment opportunities. The India Skills Report (2021) has reported,

> A decline in the employable youth population is not a massive hit for one of the fastest growing economies. If more youth are not trained and job-ready, the industry demand will not be met, but the competition is severe in India's saturated talent pool.

This skills gap prevailing between the syllabus and pedagogical approach of higher education institutes, and the industrial prerequisites, affects the employment opportunities and the self-development of the engineering graduates.

The industrial needs substitute to change with regard to the constant transformations in the world entailing the evolution of skills demands. This exhorts the students to be empowered with the adequate skills set for promoting and preparing themselves to successfully encounter the imminent challenges in their career. The India Skills Report (2021) expounds,

> Among the test takers, 45.9% were found highly employable. With a population of 1.38 billion citizens, more than half the population is estimated to be below 26 years of age. This statistic is a reflection of

DOI: 10.4324/9781003268529-3

the competition for employment in every industry. Although the formal and informal sectors offer various opportunities for graduates, the skills required are constantly evolving. Along with the rapid advancements in technology, new jobs are making way to the market.

This represents the colossal competitiveness in the global market, as a result, engendering the employers to recruit the skillful candidates for achieving successful productive outcomes of the organizations. It is construed from these evidences that the lacunae in the skills set of engineering students is in the dire need to be addressed by the higher education institutions, by exploring innovative pedagogical methods in the engineering curriculum. On the account of addressing this necessity, the present chapter investigates the facilitation of 21st Century skills through learning strategies in the classroom of engineering students.

21st Century Skills

The essence of 21st Century skills are the calibre and faculties that influence and determine the success factor of engineering students' career in this 21st Century digital informative era. The 21st Century skills are defined as aptitudes and attitudes required for employees (Mekala et al., 2020). Voogt and Roblin refer 21st Century skills as, "an overarching concept for the knowledge, skills and dispositions that citizens need to be able to contribute to the knowledge society" (as cited in Joynes et al., 2019). "21st Century Skills are psycho-social competence skills set necessary to thrive in the constantly changing global workplace" (Harishree & Mekala, 2020). These definitions depict the impact of 21st Century skills in delivering the skill-based vital needs of industrial sectors. Correspondingly, these revelations elucidate the exorbitant liability of higher education institutions to promote the 21st Century skills of engineering students with regard to fulfilling the workplace demands. The India Skills Report (2021) has posited, "The inclusion of technology assets across industries has created a high demand for professionals with a knack for learning new skills and industry specific certifications". This exhibits the need for fostering the engineering students with trailblazing skills set to achieve success in the 21st Century workplace. On this account, the present chapter propounds the implementation of 21st Century skills in the classroom of engineering students to equip them towards cumulative career prospects. Further, the chapter has focused on the employment of 21st Century skills framework developed by Partnership for 21st Century Skills (P21) considering its wide-reaching success as opined by Mekala et al. (2020), "This (P21) framework has been adopted in many countries across the globe and been successful in improving the students' skills set". P21 has consolidated the necessary skills set for the students in the 21st Century and has classified

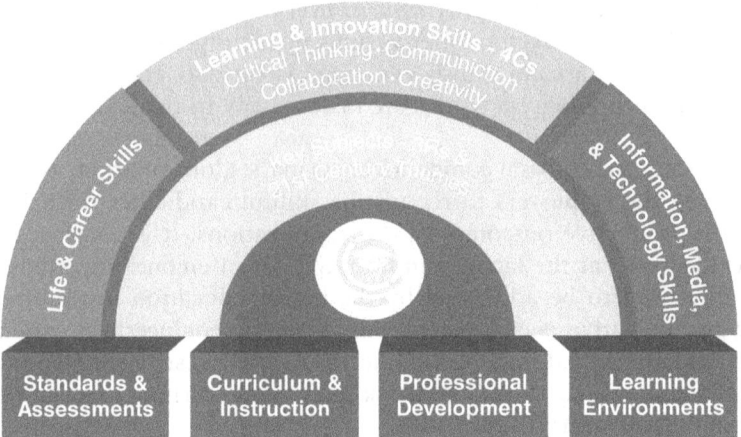

Figure 2.1 P21 Framework of 21st Century Skills Set.

Source: P21, 2019.

21st Century skills into learning and innovation skills, information, media and technology skills, and life and career skills as shown in Figure 2.1. The present chapter confines its focus towards the promotion of learning and innovation skills encompassing critical thinking, communication, collaboration, and creativity (4Cs) in the classroom of engineering students. Further, the preponderance of 4Cs for engineering students is emphasized by Samavedham and Ragupathi (2008) as, "In today's global economy, sharing of information, collaborative team work, innovative thinking, problem-solving and decision-making are key competencies necessary for an engineer". Moreover, there arises a question on the methodology to facilitate these 21st Century skills in the engineering curriculum. There are many pedagogical approaches to present the students with the cognizance of 21st Century skills but there isn't any innovative methodology to make the students sustain their knowledge on 21st Century skills and develop autonomy in learning. However, the present chapter recommends strategy-based approach for effectuating the students to not only be aware of the skills set, but also to be able to exert it in their real-life, and to be able to evolve into a self-directed learner in the long run. Moreover, the chapter proclaims the effectiveness of strategy-based approach to instigate the 21st Century skills set of the engineering students.

4Cs for Engineering Students

The proclamations by experts and skills reports confirm the insufficiency of traditional engineering curriculum to prepare engineering students for the global workplace. Since, job market is more competitive, engineers are expected to demonstrate their skills along with their subject knowledge. Albeit being many training programmes offered by different organizations and companies to prepare the engineers for job market, there is an exigency for the skill-based curriculum in the engineering institutions. Nevertheless, the skills gap of the engineering students creates a necessity for the inculcation of skill-based panacea in the engineering curriculum to hone the students' competency with 21st Century skills for being successful in the competitive global workforce. According to the AMA Critical Skills Survey (2012), employers (managers and executives) assess job applicants on the basis of the 4Cs in the process of hiring. In concord with this, the 4Cs of 21st Century skills are identified to be of utmost importance for today's engineers to promote productive outcomes in the workplace. Critical thinking, communication, collaboration, and creativity (4Cs) configure the foundation platform for engineers to play multidisciplinary roles in their workplace environment. This evinces the significance of 4Cs in augmenting the success of the students' future career.

Critical Thinking

Critical thinking skills embrace the cognitive operations involved in higher order thinking skills say remembering, understanding, applying, analysing, revising, and creating. As cited in Xu (2011),

> Critical thinking includes not only critical thinking skills (containing both a process of thinking and thinking ability), involving analysis, interpretation, inference, explanation, evaluation, and self-regulation but also critical thinking dispositions including clarity, accuracy, precision, consistency, relevance, sound evidence, good reasons, depth, breadth, fairness.
>
> (Scriven & Paul)

Pardede (2020) expounds critical thinking as intricate mental process entailing a myriad and multidimensional cognitive ability. Engineering students of the 21st Century should be able to analyse, synthesize, and evaluate every information they learn. In the digital era of the 21st Century, multitudinous information are available on online platforms, but the source of authentication has to be decided by the students. This evinces that the students should be equipped to think, generate, analyse, and decide by themselves on the source of authentication of the online resources. Cooney et al. (2008) insist that the critical thinking skills is the ability to analyse complex issues, synthesize information, evaluate logic and validity, solve challenging problems, and generate and explore new ideas. Besides, Lunt and Helps (2001) have elucidated the significant traits of critical thinking skill as indicated in Table 2.1. (As cited in Cooney et al., 2008.)

Table 2.1 Traits Important to Critical Thinking (as Cited in Cooney et al., 2008)

Revise a previously held view to account for new information
Determine associations between similar ideas, objects, and situations
Analyse knowledge within a given domain and context
Be willing to evaluate an argument or proposition posed by an authority
Generate solutions to problems
Identify the most significant variables involved in a problem
Relate what kind of evidence will support a thesis or hypothesis
Produce an argument that is internally consistent
Clearly identify central issues and problems to be investigated or hypothesis
  to be tested
Determine whether the conclusions drawn are logically consistent with, and
  adequately supported by, data or accepted information
Distinguish between relevant and irrelevant information
Assemble facts to determine the validity of an argument
Explain ideas with reasonable clarity
Draw sound inferences from the information found or given
Critically reflect on and analyse all information presented

The predominance of critical thinking skills for engineering students is propounded by Ivleva (2016) as,

> Critical thinking for engineering students, future engineers, is supposed to be more important as these specialists work in the areas of telecommunication, information and communication technologies, mechatronics and robotics, physics, machine building and other engineering areas which require accuracy and fidelity, clearness of actions, practical thinking, attentiveness, and systematization.

These provide outline on the endeavours that engineering students should be able to venture competently in their future career. This asseverates the desideratum of critical thinking skills for the engineering graduates to achieve progressive outcomes in their enterprise.

Communication

The term 'Communication' has become a widely expressed platitude in terms of enticing extensive learners towards the language courses. Irrespective of embellishing the language learning courses and engrossing majority of the syllabus, communication skills are identified to be deficient in the employees at the workplace scenario. Keane and Gibson have emphasized, "Communication skills are a regular feature of an engineer's job in industry; some graduates employed in industry have identified that education in communication skills needs to be improved given the demands encountered in industry" (as cited in Riemer, 2007). Communication skills are cardinal skills not only for engineers but also for other professionals to convey their

Table 2.2 Communication Skills (Nutman, 1987)

Information Retrieval	• A comprehensive knowledge of the availability and method of retrieval of academic literature: texts, journals, abstracts, indexes, and industrial publications. • The ability to locate and communicate with other academics working in the same area. The ability to locate and communicate with engineers working in the area in industry.
Comprehension and Summarizing	The ability to understand and summarize the main features of an article and present these in their own words.
Written Communication	The ability to design and present in writing a substantial report on a productive engineering problem.
Oral Presentation	The ability and confidence to present their findings to their peers and academic staff in a manner that demonstrates understanding and interest.

thoughts to the receivers with brevity and clarity. Though, communication skill is an indispensable component in the education of engineering students to prepare them for their future careers (Riemer, 2007), the pedagogical focus of the skill is of rudimentary basis, impeding the students from learning the multifaceted in-depth constituents of communication skill. A few principal communication skills expected of engineering graduates are tabulated by Nutman (1987) as presented in Table 2.2.

Nestsiarovich (2020) has enumerated the role of communication in the workplace as,

> engineering communication is understood to include: presentations; project discussions at meetings; written communication between engineers or engineering departments, such as feedback and emails; vertical communication, such as that between superiors and subordinates, and informal communication at the workplace; and non-verbal communication.

"Engineers equipped with effective communications skills are assets to employers who are searching for employees able to play multiple roles to promote business and their company" (Lenard & Pintaric, 2018). Miceli has enunciated that the increase in the instantaneous communications with regard to the viral usage of social media, the ascendancy of concise, effective, conveyance of ideas has become significant. Further, she has postulated, "engineers' lack of communication skills has contributed to a widespread undervaluing of engineering as a discipline" (as cited in Linvill et al., 2019). These studies stress the predominance and necessity of communication skills for the engineering students, who are in the desperate need to be facilitated with these skills in the engineering classroom for assisting them in their recruitment process and in their workplace environment.

Collaboration

Collaboration is constituted as a social and teamwork skill, traditionally practised in the classrooms with debate and group discussion activities. Dede (2009) refers collaboration as, "the practice of working together to achieve a common goal. It is an increasingly important educational outcome because organizations and businesses have increasingly moved to a team-based work environment" (as cited in Pardede, 2020). Panitz has proclaimed that collaboration is a philosophy of interaction and personal lifestyle to accomplish an end product or goal (as cited in Gol and Nafalski, 2007). Hamilton et al. have asserted, "As global competition increases, organizations need to perform smarter, faster, and more efficiently. This requires embedding collaborative technologies deep into processes and incentivizing collaborative behaviors—ultimately transforming the way organizations—even classroom learning situations—turn knowledge into action" (as cited in Marra et al., 2016). Gol and Nafalski (2007) present shared goals, teamwork, individual responsibility, and self-discipline as the tenets of collaborative learning. Vila et al. (2017) have insisted, "collaboration is required in situations where several persons have to make decisions. Decisions are driven by preferences and entail socially constructed knowledge derived from human opinions and interactions". Accordingly, engineers are expected to work with diverse teams with persuasiveness and efficient contribution to the team, for effective collaborative outcome. Pazos et al. (2016) have asserted, "working in teams requires that students learn how to interact with each other while sharing and processing information in a collaborative learning environment". In this line, they are expected to possess the collaboration skills necessary to take responsibility, discuss in groups, present their ideas with their team members productively. Vogler et al. have stated, "the most significant benefit of teaching teamwork skills in an academic setting is their potential to transfer to the workplace" (cited in Pazos et al., 2020). These findings reveal the necessity of engineering students to possess collaboration skill for presenting their perceptions across their team members and being able to produce an effective outcome in their workplace.

Creativity

Creativity is an innate skill with ingrained fascination, and inspiration to explore the unexplored possibilities. Plucker et al. define creativity as, "Creativity is the interaction among aptitude, process, and the environment by which an individual or group produces a perceptible product that is both novel and useful as defined within a social context" (as cited in Tep et al., 2021). Hence, novelty and inventiveness are regarded as the most vital aspects of creativity skill, which encompass brainstorming, convergent, and divergent thinking skills. Industries rely on creativity to think out of box, to question assumptions and also to find solutions to problems. "With the advent of knowledge-based economies in the 21st century, creativity has become a crucial issue for human-resource management in both academic research and industry" (Chen and Hsu, 2006). According to the AMA Critical Skills Survey (2012), 91.6% of the employers give importance to creativity. Engineers in working place are expected to be creative to think discretely with shedload of ideas for producing ingenious products or untangling the complications of the problems. As cited in Nakano and Wechsler (2018), "The search for creative professionals who can innovate – that is, individuals who stand out for their mastery of efficient strategies to address new problems and solve them successfully – has been emphasized by different types of organizations" (Cropley, 2005). Besides, "The attributes students most frequently associated with creativity included novelty, fluency, open-mindedness, unconventionality, synthesis, insightfulness, and attitude" (Richards, 1998). Taylor has classified creativity as shown in Figure 2.2 (as cited in Liu & Schonwetter, 2004).

Martin (1991) has emphasized, "engineers need a creative mind to meet the advancing goal of the engineering profession – to design new products or systems and improve existing ones for the benefit of humankind" (as cited in Liu & Schonwetter, 2004). Prahalad and Ramaswamy have expounded, "The ability to engage in a creative process to define or solve a problem or design a novel artifact is essential to engineering as a

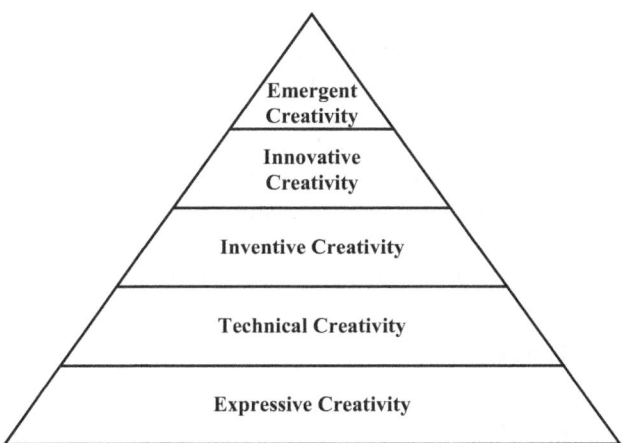

Figure 2.2 Taylor's Hierarchy of Creativity.

(Figure Based on Liu & Schonwetter, 2004).

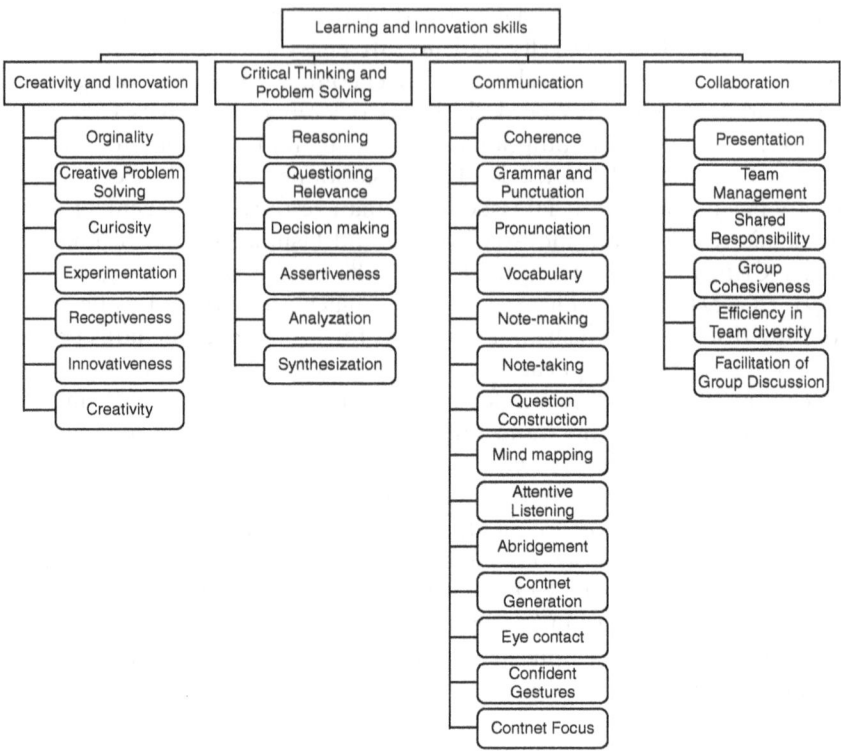

Figure 2.3 Tenets of Learning and Innovation Skills.

Source: Author.

profession, and especially to future engineers" (as cited in Daly et al., 2014). These studies emphasize the importance of facilitating the engineering students with creativity skill for their successful ventures of future career prospects.

These revelations exhibit the desideratum of these skills (4Cs) in capacitating the engineering students with the cardinal workplace skills. The present chapter emphasizes the tenets of 4Cs customized by Mekala et al. (2020) as displayed in Figure 2.3 for incorporating in the pedagogical approaches of engineering classroom in integration with learning strategies.

Learning Strategies

In the 21st Century, students should be aware, and be able to systematize and regulate their skills competence. They should be able to critically analyse, monitor, and evaluate their learning progress for it promotes autonomy in learning. Students will eventually require autonomy to self-motivate,

Table 2.3 Classification of Learning Strategies (as Cited in Vlckova et al., 2013)

Groups	Subgroups	Learning Strategy
Metacognitive	Planning	Advance organizers, directed attention, selective attention, self-management, functional planning
	Monitoring	Self-monitoring
	Evaluation	Self-evaluation
Cognitive	—	Repetition, resourcing, translation, grouping, note taking, summarizing, deduction, recombination, translation, imagery and auditory representation, keyword method, contextualization, elaboration, transfer, inferencing.

self-assess, and self-regulate themselves at their workplace. The students would ultimately transfer their knowledge learnt at college to their workplace. The transfer of learning, problem solving, critical thinking, self-awareness, self-efficacy, etc., are the prerequisites for the workplace as well as for self-directed learning (analysing learning needs, setting goals, taking responsibility, and regulating learning on own). These skills necessitate students to develop insight into their learning needs and attitudes, and to plan and reflect on their learning progress. This shows that the engineering students are in exigency to be facilitated with the knowledge and competence to cope with the unforeseen workplace challenges on their own, which can be effectuated through the employment of 21st Century skills in amalgamation with learning strategies.

Lenz et al. have stated learning strategies as, "an individual's approach to a task. It includes how a person thinks and acts when planning, executing, and evaluating performance on a task and its outcomes" (as cited in Schumaker & Deshler, 1992). Learning strategies are techniques or devices (Rubin, 1975) and cognitive operations (Chaudron, 2009) that the learners employ to enhance their learning process. Vlckova et al. (2013) exhibit that the learning strategies are preponderant in language acquisition and educational achievement. Further, they enumerated the wide range of classifications on learning strategies by theorists (Naiman et al., 1978; Dansereau, 1985; Weinstein & Mayer, 1986; Wenden & Rubin, 1987; O'Malley, Chamot, Stewner-Manzanares, Kupper, & Russo, 1985; Oxford, 1990; Stern, 1992; Ellis, 1990; Bimmel & Rampillon, 2000; Cohen & Weaver, 2006; etc.). The present chapter has focused on the pivotal learning strategies say cognitive and metacognitive strategies to foster the creativity, critical thinking, communication, and collaboration skills of the engineering students as presented in Table 2.3.

Cognitive Strategies

As cited in Shawer et al. (2008),

> cognitive strategies are the 'steps or mental operations used in learning or problem-solving that require direct analysis, transformation, or synthesis of learning materials in order to store, retrieve, and use knowledge' (Wenden). 'cognitive strategies involve asking questions, checking, revising, self-testing' (Riding & Rayner); 'analogy, memorization, repetition, writing things down, and inference.
>
> (Hedge)

Gamage (2003) propounds that cognition refers to the systematic mechanism of brain to process and retrieve the information. Besides, Prokop opines that the cognitive strategies are the processes that involve the task at hand and the manner of linguistic information management (as cited in Setiyadi, 2001). In brief, cognitive strategies are those that equip and guide the students to hone their cognition strategically for achieving constructive outcome. Such cognitive strategies are classified on numerous factors such as source-based: self-created, externally imposed; orientation-based: person, task; purpose-based: learning, performance; scope-based: task-specific, task-general (Singer and Chen, 1994); deep level: deduction, recombination, keywords; and surface level: repetition, note-taking, auditory representation, resourcing (Setiyadi, 2001). The detailed explanation of cognitive strategies (White, 1993) is presented in Table 2.4.

Table 2.4 Cognitive Strategy Descriptions and 4Cs

Cognitive Strategy	Description	Skills
Repetition	Repeating a chunk of language (a word or phrase) in the course of performing a language task.	Communication
Resourcing	Using target language reference materials such as dictionaries, encyclopedias, or textbooks.	Communication
Grouping	Ordering, classifying, or labelling material used in a language task based on common attributes.	Communication and Collaboration
Note Taking	Writing down key words and concepts in abbreviated verbal, graphic, or numerical form to assist performance of a language task.	Communication
Deduction	Consciously applying learned or self-developed rules to produce or understand the second language.	Critical Thinking and Collaboration

(Continued)

Table 2.4 (Continued)

Cognitive Strategy	Description	Skills
Substitution	Selecting alternative approaches, revised plans, or different words or phrases to accomplish a language task.	Critical Thinking and Collaboration
Imagery elaboration	Using mental or actual pictures to represent information.	Creativity and Critical Thinking
Visualization	Using visual stimuli to clarify meaning.	Critical Thinking, Creativity, and Collaboration
Word Elaboration	Relating new information to prior knowledge gained from experience in the world.	Creativity, Critical Thinking and Collaboration
Between Parts Elaboration	Relating parts of the task to each other.	Creativity and Critical Thinking
Contextualization	Placing a word or phrase in a meaningful language sequence.	Communication and Creativity
Summarization	Making a mental or written summary of language and information presented in a task.	Communication, Creativity, and Collaboration
Translation – To English	Using LI as a base for understanding L2.	Communication
Translation – From English	Using LI as a base for producing L2.	Communication
Inferencing	Using available information to guess the meanings or usage of unfamiliar language items associated with a language task.	Critical Thinking, Creativity, and Collaboration
Transfer	Using previously acquired linguistic knowledge to facilitate a language task.	Creativity, Critical Thinking, and Collaboration
Rehearsal	Rehearsing the language needed, with attention to meaning, for an oral or written task.	Communication and Critical Thinking

The efficacious contribution of cognitive strategies in stimulating the engineering students' cognizance and cogitation on 4Cs is apparent from Table 2.4. It is distinct that these sub strategies of cognitive strategy can be recommended to be implemented in the 21st-Century classroom of engineering students in integration with the pedagogical methods/activities for ensuring the students' competence to compete in the world of work. These cognitive strategies play a cardinal role in enhancing the students' workplace skills that are demanded of them for the productive outcome of organizations.

Metacognitive Strategies

The cognizance of one's own cognition is metacognition. According to Flavell, "Metacognition is a knowledge or cognitive activity that takes as its object, or regulates, any aspect of cognitive activity" (as cited in Kim, 2016). Devine has stated, "Metacognition is a form of cognition which includes active control over cognitive procedure" (as cited in Nazarieh, 2016). Metacognition encompasses two components say metacognitive knowledge and metacognitive experiences. Flavell (1979) has defined metacognitive knowledge as the awareness of one's tasks, goals, actions, and experiences and has classified it into person, task, and strategy knowledge. Metacognitive experience refers to the cognitive endeavours of an individual. In addition, Brown (1987) has proposed declarative, procedural, and conditional knowledge. Declarative knowledge is the knowledge of an individual about him/herself, while procedural knowledge is the consciousness of the individual's skills and conditional knowledge is the awareness to apply the skills and strategies at appropriate juncture. Akturk and Sahin (2011) have expressed, "Students with advanced metacognitive skills may monitor their own learning, express their opinions about the information, update their knowledge and develop and implement new learning strategies to learn more". Metacognition empowers the students to be self-directed learners. It promotes their strategic and reflective thinking. The metacognitive strategies: planning, monitoring, and evaluating capacitate the students to progress effectively in their learning process. Metacognitive strategies are used by learners to regulate their learning. It helps in planning, arranging, and evaluating the learned information. It also supports selective attention, self-management, self-monitoring, identifying problem, and self-evaluation. The comprehensive explanation of metacognitive strategies (White, 1993) is represented in Table 2.5.

Table 2.5 reflects the crucial role of metacognitive strategies to facilitate the engineering students in the process of skills acquisition if integrated in

Table 2.5 Metacognitive Strategy Descriptions and 4Cs

Metacognitive Strategy	Description	Skills
Advance Organization	Previewing the organizing concept or principle of an anticipated learning task.	Communication, Critical Thinking, and Collaboration
Selective Attention	Deciding in advance to attend to specific aspects of input, often by scanning for key words, concepts, and/or linguistic markers.	Communication and Collaboration
Directed Attention	Deciding in advance to attend in general to a learning task and to ignore irrelevant distractors.	Communication and Critical Thinking

(Continued)

Table 2.5 (Continued)

Metacognitive Strategy	Description	Skills
Delayed Production	Consciously deciding to postpone speaking to learn initially through listening comprehension.	Communication, Critical Thinking, and Collaboration
Self-Management	Understanding the conditions that help one successfully accomplish language tasks and arranging for the presence of those conditions.	Critical Thinking and Creativity
Problem Identification	Explicitly identifying the central point needing resolution in a task or identifying an aspect of the task that hinders its successful completion.	Critical Thinking, Creativity, and Collaboration
Self-Monitoring	Checking one's comprehension during listening or reading or checking the accuracy and/or appropriateness or one's oral or written production while it is taking place.	Communication and Critical Thinking
Self-Evaluation	Checking the outcome of one's own language learning against an internal measure of completeness and accuracy.	Critical Thinking
Prioritizing	Prioritizing learning according to one's personal needs and/or wants.	Critical Thinking and Creativity
Revision	Systematically reviewing in order to aid long-term retention.	Critical Thinking and Collaboration

their classroom. It is exhibited that the respective sub strategies of metacognitive strategy can enable the students with the respective 21st Century skill set that empowers the students to produce constructive outcome in the workplace as well as to augment their independent learning attitude.

While cognitive strategies assist the students to perform better in the respective tasks, metacognitive guides them to analyse and reflect on the progress and performance of the task. In line with this, while cognitive strategies promote the engineering students with 21st Century skills, the metacognitive strategies enhance the self-regulation of the students to establish their autonomy in the process of learning.

Implementation

Cognition and metacognition play pivotal role in effectuating and regulating the conscious thought process of the students, which is the essential requisite for improving the students' skills set. In terms of communication, the students would have to observe, interpret, perceive, decipher, take and make notes, write with clarity, speak fluently, summarize, etc. These tenets require them to instigate their consciousness, which is possible through the employment of cognitive and metacognitive strategies. Critical thinking and creativity engender the students to analyse, synthesize, generate new ideas, think radically, plan, evaluate, reflect, etc., necessitating reflective and strategic thinking. Besides, the collaboration skill incites the students to mingle with diverse people, state the opinions distinctly, select the appropriate and relevant ideas generated through discussion with peers, etc., entailing social thinking and deductive reasoning. Consequently, the engineering graduates are expected to be adept to infer, reflect, plan, organize, set goals, manage time, lead the team members, solve problems, produce innovative ideas/techniques, communicate information, explicit abstraction, transfer knowledge, etc., among their colleagues, managers, higher officials, and clients at their workplace. Such preponderant skill sets are recommended to be facilitated to the students by employing cognitive and metacognitive strategies in the engineering classroom. The teachers should instill the students with the awareness of these strategies and also with the knowledge to appropriately employ these strategies in the allocated tasks on skills set. The syllabus should be designed with the activities/tasks that stimulate the students' communication, critical thinking, creativity, and collaboration skills. However, it is the teachers' responsibility to integrate these skill-based activities with the learning strategies for constructive achievement in augmenting students' skills set.

Teachers should instruct and guide the students at the beginning of each task on the strategies that can be employed for respective tasks to expedite their task performance and skills acquisition. Explicit instruction on the tasks and strategies will provide the students with clarity on the process and completion of the tasks. This in turn, would promote their task, person, and strategic knowledge to plan and perform their task effectively. The sub strategies in Tables 2.4 and 2.5 can be employed by the engineering students in skill-based activities facilitated in the classroom, under the guidance of the teacher for s/he provides the students with the exposure to utilize appropriate strategies at appropriate tasks. The absence of this knowledge would not lead the students towards the accomplishment of effective skills competence. In spite of providing the students with the knowledge of the strategies and tasks, the teachers should provide the students with the insight on the outcome and usefulness of the task at their future career and life. This strategy-based approach towards the implementation of skill-based activities in the engineering classroom would motivate the students' erudition and competence, eventually procuring adequacy in 21st Century skills set and self-directed learning.

Classroom-Based Activities

The effective integration of learning strategies and 21st Century skills in the engineering classroom is determined by the teachers' effective facilitation of these strategies and skills to the students of engineering by employing ingenious pedagogical methods. The present chapter guides the teachers by recommending few activities suggested by researchers and theorists for the implementation of skill-based activities through task-based approach in the engineering classroom.

Activities say 'Particular Virtues', 'Per Mission', 'Perspicacious Perspective', 'Turnarounds', 'Peerless Recognition', '(Finger) Food of Thought', 'Try! Umph!', 'Left is Right. So is Right', 'Ms. Matches and Mr. Matches', 'Cre8-GetN2It', 'Think Outside the Locks', 'A Kin to Kinesthesia', 'Low and High Logos', 'Blues on Parade', 'Scrambled Pegs', and 'Ban Banalogies' have been proclaimed by Caroselli (2009) for enhancing the learners' creative thinking skills.

Activities such as 'Time Capsule', 'Mind Squeeze', 'Trivia Trackdown', 'Workplay', 'Making a Menu', 'Recycled Words', 'What Am I', 'Arctic Facts', 'Antarctic Facts', 'What's Up & What's Down?', and 'Transformations' are expounded by Rozakis (1998) for effectuating the critical thinking skills of the learners.

Garber (2008) posits that the activities like 'Reading Test', 'The Longest Line', 'Color Block', 'Following Directions', 'Train Story', 'Count the Ss', 'Communications Shutdowns', 'Repeat the Questions', 'Quick Answers', 'Rephrasing Exercise', 'Wedding Story', 'Who done it', 'Picture Puzzle', 'Creative Fairy Tale', 'Rumors', 'The Show Store', 'Two Mouse Sitting on a Log', 'Communications Styles', 'Oh', 'Word Games', and 'Trading Cards' can be facilitated to enhance the communication skills of the learners.

Osterholt and Barratt (2012) recommend 'Teaching to Learn', 'Understanding Controversy', 'Using Group Strength', 'Understanding Text Material', and 'Meaningful Mingling' activities for promoting the students' collaboration skill.

However, the administration of these activities in the classroom requires appropriate planning that the teachers should enhance students' cognizance on the strategies embedded in each task to instigate their cognition and to expedite their skill acquisition. After each task, the students should be encouraged to evaluate and reflect on their self and peers' performance, as it will aid them towards self-directed learning.

Conclusion

The skills gap persisting between the students' skills set and the skills demanded by industries creates an exigency for the higher education institutes to modify the engineering curriculum in line with the transformations in the 21st-Century workplace. In order to address this need, the present chapter propounds the integration of learning strategies: cognitive and metacognitive strategies and the 21st Century skills: communication, collaboration, creativity, and critical thinking skills (4Cs) with the engineering curriculum. Further, it emphasizes the necessity of learning strategies in augmenting the skills set of engineering students to meet the requirements of their future workplace. The chapter asserts that the learning strategies expedite the students in the process of acquisition of 21st Century skills. Moreover, it has recommended activities framed by theorists and researchers to promote the 4Cs of the engineering students by aiding the teachers in their task-based pedagogical approach. However, the present chapter has limited its scope to the enhancement of learning and innovation skills of the engineering students from theoretical perspectives that the future studies can explore on the possibilities of these theoretical implications through an experimental study.

References

Agarwal, P. (2006). *Higher education in India: The need for change* (ICRIER Working Paper No. 180). Indian Council for Research on International Economic Relations. http://www.icrier.org/pdf/ICRIER_WP180__Higher_Education_in_India_.pdf

Akturk, A. O., & Sahin, I. (2011). Literature review on metacognition and its measurement. *Procedia Social and Behavioral Sciences, 15*, 3731–3736.

American Management Association. (2012). *AMA 2012 Critical Skills Survey.* https://www.amanet.org/assets/1/6/2012-critical-skills-survey.pdf.

Brown, A. L. (1987). Metacognition, executive control, self-regulation and other more mysterious mechanism. In F. E. Weinert & R. H. Kluwe (Eds.), *Metacognition, motivation, and understanding* (pp. 65–116). Hillsdale, NJ: Lawrence Erlbaum Associates.

Caroselli, M. (2009). *50 activities for developing critical thinking skills.* Massachusetts: HRD Press.

Chaudron, C. (2009). *Second language classrooms: Research on teaching and learning.* New Delhi: Cambridge University Press.

Chen, C., & Hsu, K. (2006). Creativity of engineering students as perceived by faculty: A case study. *International Journal of Engineering Education, 22*(2), 264–272.

Cooney, E., Alfrey, K., & Owens, S. (2008). AC 2008-1110: Critical thinking in engineering and technology education: A review. *2008 Annual Conference & Exposition: Improving Technical Understanding of All Americans*, Pittsburgh, Pennsylvania. American Society for Engineering Education.

Daly, Shanna et al. (2014). Teaching creativity in engineering courses. *Journal of Engineering Education, 103*(3), 417–449.

Dede, C. (2009). Comparing frameworks for "21st century skills". In J. Bellanca & R. Brandt (Eds.) *21st century skills: Rethinking how students learn* (pp. 51–76). Bloomington, IN: Solution Tree Press.

Flavell, John H. (1979). Metacognition and cognitive monitoring: A new area of cognitive – developmental inquiry. *American Psychologist, 34*(10), 906–911.

Gamage, G. H. (2003). Issues in strategy classifications in language learning: A framework for Kanji learning strategy research. *ASAA e-Journal of Asian Linguistics & Language Teaching, 5*, 1–14.

Garber, P. R. (2008). *50 communications activities icebreakers, and exercises.* Massachusetts: HRD Press.

Gaurav, J. (2020). *Skill gap analysis of civil engineering sector in India: Skills needed to succeed in job market.* Chennai: Notion Press.

Gol, O., & Nafalski, A. (2007). Collaborative learning in engineering education. *Global Journal of Engineering Education, 11*(2), 173–180.

Harishree, C., & Mekala, S. (2020). Need for 21st century skills education for teachers. *Roots International Journal of Multidisciplinary Researches, 7*(1), 45–52.

India Skills Report. (2021). *Key insights into the post-COVID landscape of talent demand and supply in India.* Haryana: Wheebox. Retrived from https://wheebox.com/india-skills-report.htm#

Ivleva, N. V. (2016). Teaching critical thinking to engineering students through reading profession-oriented texts. *IOP Conference Series: Materials Science and Engineering, 155,* 012022. DOI: 10.1088/1757-899X/155/1/012022

Joynes, C., Rossignoli, S., & Fenyiwa Amonoo-Kuofi, E. (2019). *21st century skills: Evidence of issues in definition, demand and delivery for development contexts* (K4D Helpdesk Report). Brighton, UK: Institute of Development Studies.

Kim, M. M. J. (2016). Writing about writing: Qualities of metacognitive L2 writing reflections. *Second Language Studies, 34,* 1–54.

Richards, L. G. (1998). Stimulating creativity: teaching engineers to be innovators. In *FIE '98. 28th annual frontiers in education conference: Moving from 'Teacher-Centered' to 'Learner-Centered' education. Conference proceedings (Cat. No. 98CH36214)* (pp. 1034–1039). Tempe, AZ: USA. DOI: 10.1109/FIE.1998.738551.

Lenard, D. B., & Pintaric, L. (2018). Communication skills as a prerequisite for the 21st century engineer. *ELT Vibes: International E-Journal for Research in ELT, 4*(2), 11–45.

Linvill, N. Darren, et al. (2019). Engineering identity and communication outcomes: Comparing integrated engineering and traditional public-speaking courses. *Communication Education, 68*(3), 308–327. DOI: 10.1080/03634523.2019.1608367

Liu, Z., & Schonwetter, D. J. (2004). Teaching creativity in engineering. *International Journal of Engineering Education, 20*(5), 801–808.

Marra, Rose, et al. (2016). Beyond "group work": An integrated approach to support collaboration in engineering education. *International Journal of STEM Education, 3*(17), 1–15.

Mekala, S., Harishree, C., & Geetha, R. (2020). Fostering 21st century skills of the students of engineering and technology. *The Journal of Engineering Education Transformations, 34*(1), 75–88.

Nakano, T. C., & Wechsler, S. M. (2018). Creativity and innovation: Skills for the 21st century. *Estudos de Psicologia* (Campinas)*, 35*(3), 237–246. DOI: 10.1590/1982-02752018000300002.

Nazarieh, M. (2016). A brief history of metacognition and principles of metacognitive instruction in learning. *BEST: Journal of Humanities, Arts, Medicine and Sciences (BEST: JHAMS), 2,* 61–64.

Nestsiarovich, K. (2020). *Communication in engineering teams: Personal interactions and role assignment.* [Unpublished Doctoral Thesis]. University of Canterbury.

Nutman, P. N. S. (1987) Communication skills for engineering students: An integrative approach, *European Journal of Engineering Education, 12*(4), 367–375, DOI: 10.1080/03043798708939383.

Osterholt, D. A., & Barratt, K. (2012). Ideas for practice: A collaborative look to the classroom. *Journal of Developmental Education, 36*(2), 22–44.

Pardede, P. (2020). Integrating the 4Cs into EFL integrated skills learning. *Journal of English Teaching, 6*(1), 71–85.

Partnership for 21st Century Learning: A Network of Battle for Kids. (2019). *Framework for 21st Century learning definitions.* http://static.battelleforkids.org/documents/p21/P21_Framework_DefinitionsBFK.pdf

Pazos, Pilar et al. (2016, June 26–29). Developing critical collaboration skills in engineering students: Results from an empirical study. Paper presented at *ASEE's 123rd annual conference & exposition*, New Orleans, LA.

Pazos, Pilar et al. (2020, June 22–26). Enhancing teamwork skills through an engineering service learning collaboration. *Paper presented at ASEE's virtual conference, Virginia: Old Dominion University.*

Riemer, M. J. (2007). Communication skills for the 21st century engineer. *Global Journal of Engineering Education, 11*(1), 89–100.

Rozakis, L. (1998). *81 fresh & fun critical-thinking activities: Engaging activities and reproducibles to develop kids' higher-level thinking skills.* The U.S.A: Scholastic Teaching Resources.

Rubin, J. (1975). What the 'good language learner' can teach us. *TESOL Quarterly, 9*, 41–51.

Samavedham, L., & Ragupathi, K. (2008). Facilitating 21st century skills in engineering students. *The Journal of Engineering Education, 26* (1), 37–49.

Schumaker, J. B., & Deshler, D. D. (1992). Validation of learning strategy interventions for students with LD: Results of a programmatic research effort. In Y. L. Wong (Ed.), *Contemporary intervention research in learning disabilities: An international perspective.* New York: Springer-Verlag.

Setiyadi, Ag. B. (2001). Language learning strategies: Classification and pedagogical implication. *TEFLIN Journal, 12*(1), 15–28.

Shawer, Saad et al. (2008). Student cognitive and affective development in the context of classroom-level curriculum development. *Journal of the Scholarship of Teaching and Learning, 8*, 1–28.

Singer, R. N., & Chen, D. (1994). A classification scheme for cognitive strategies: Implications for learning and teaching psychomotor skills. *Research Quarterly for Exercise and Sport, 65*(2), 143–151, DOI: 10.1080/02701367.1994.10607609

Tep, P., Maneewan, S., & Chuathong, S. (2021). Psychometric examination of Runco ideational behaviour scale: Thai adaptation. *Psicologia: Reflexao e Critica, 34*(4), 1–11.

Vila, C. et al. (2017). Project-based collaborative engineering learning to develop industry 4.0 skills within a PLM framework. *Procedia Manufacturing, 13*, 1269–1276.

Vlčkova, Kateřina et al. (2013). Classification theories of foreign language learning strategies: An exploratory analysis. *Studia Paedagogica, 18*, 93–113.

White, C. J. (1993). *Metacognitive, cognitive, social and affective strategy use in foreign language learning: A comparative study.* [Unpublished Doctoral Thesis]. Massey University, New Zealand.

Xu, Jun. (2011). The application of critical thinking in teaching English reading. *Theory and Practice in Language Studies, 1*(2), 136–141.

Part II

Communication Skills

Chapter 3

Bridging the Skills Gap in Technical Education

C. Shanmuga Priya

The expectations of the industry in recruiting skilled incumbents remain a daunting question in the Indian manufacturing sectors. Employers strive hard to fill the thriving job openings in their organizations despite a massive number of unemployed engineers. Employers observe a lack of critical thinking and communication skills in today's workforce and propound their employees' ineptitude to meet the expectations of the industry to communicate the business strategies in the global market to be the reason for the skills gap. The skills deficiencies in manufacturing sectors, for example, lack of enterprising skills such as generating, comprehending, and communicating the content for external operations, sustaining market perception and brand building, and functioning as a curator for press and media significantly impede their ability to expand the operations and improve productivity. The American Society for Training and Development (ASTD) defines the skills gap as "a significant gap between an organization's current capabilities and the skills it needs to achieve its goals" (Bridging the Skills Gap, 2012, p. 4). The Computer and Technology Industry Association (CompTIA) (2012) has presented the views of 93% of the employers affirming the skills gap in their organizations. Career Builders annual survey (2017) of employers in the US has predicted that 69% of the companies would struggle to hire graduates adequately over the coming years and also has identified that almost one in five employers believe the failure of academic institutions in providing adequate workplace preparation. Accenture reports that 75% of manufacturers are experiencing a moderate to severe shortage of talent. The skills of the candidates against the availability of the jobs in the market and their work readiness offer a new perspective to the academicians, employers, government, and the economy as a whole. The chapter substantiates the role and significance of communication skills in the career of engineers, the need for English language proficiency (ELP) at the workplace and the expectations of the employers pertaining to the level of employees' communication skills. The existing skills gap in engineering enterprises is highlighted through the review of literature and statistical analysis of the data. A few recommendations are suggested to bridge the skills gap based on the statistical evidences and observations.

DOI: 10.4324/9781003268529-5

Studies on Skills Gap

A survey conducted by World Economic Forum [as cited in Aring, 2012] on employers' perceptions of skills gap indicates that the employers all over the world consider talent or skills gap as one of their top five concerns. Manpower Group's 2011 [as cited in Aring, 2012] survey denotes that 34% of employers globally report that the lack of available talent causes them difficulties in filling positions. The survey also enumerates the percentage of employers reiterating the difficulties due to skills gap in the countries such as the US with 52%, Germany with 40%, India with 67%, and Japan with 80%. The World Bank Enterprise Survey [as cited in Aring, 2012] picturizes the employers' opinion that skills gap constrains their ability to compete with the progressive countries. Engineers require analytical skills and domain knowledge to apply their technical competence of core engineering involved in the process of manufacturing a product or service. The requisite proficiency in English language is indispensable to understand the instructions and respond to the situations appropriately. The National Association of Colleges and Employers (NACE, 2016) in its study with US employers, endorses that the most desired skills by the employers are the team work, problem-solving skills, communication skills, organization of ideas, and the ability to process information. Aspiring Minds' National Employability Report (2019) has stated that engineering education should be project-based. It instructs the faculty to teach the students on how to apply the engineering concepts to industry with a support strategy to address the skills gap.

In the globalized context, India is no exception in encountering the crisis of unemployability due to the lack of skill sets among the engineers. The Indian Confederation of Industries ([as cited in Aring, 2012] has confirmed that the skills gap in employability, soft skill, technical skill, and English language is extensive from the entry level to professional and upper levels and this remains one of the major constraints to the continued growth of the Indian economy. It also states that only 5% of the total Indian workforce is skilled amidst the 40% of India's population. Only 9% seems to have been engaged in the organized sector and only 5% possesses marketable skills among the 500 million workforces in India. This is echoed by the World Economic Forum's Global Talent Mobility report [as cited in Aring, 2012]. The World Bank claims that there is a profound skills gap among the Indian engineers who, according to the employers, lack all the necessary competence in soft skills. India Skills Report (2017) highlights that of all the students entering the job market across the country, hardly two of five meet the employment criteria set by the employers. Though new jobs are generated in core engineering and IT industries, there are not enough skilled persons available for the requisite cadres. Candidates with ELP other than domain knowledge are in demand in globally developed economies. The All India Council for Technical Education (AICTE) [as cited in India Skills Report, 2017] reports that out of eight lakh B.E./B.Tech. students graduating every year, only half of them get placed through campus place-ments. Employer studies from India show that among 4,50,000 engineers graduating each year in India, only 25% possess to be employable. Indian CEOs have declared that India faces a heavy shortage in skill set in spite of her large and young manpower in hand. According to The Conference Board (2008), Indian employers seem to experience a labour shortage of engineering graduates with effective communication skills in English. These studies reveal that a profound skills gap is prevailing between the education and the industrial requirement leading to the inadequacy of the graduates to grab the thriving employment opportunities.

Skills Mismatch as a Social Problem

The term 'skills mismatch' can be defined as the difference between the req-uisite workplace skills and the skills actually possessed by the employees. The existence of a skills mismatch or skill shortage is by no means as obvi-ous as often emphasized. During the 1970s, workforce skill levels have been believed to have exceeded the levels that jobs could utilize. Berg (1971) and Collins (1979) put forth that employers' inflated hiring requirements led American workers to acquire more education than they really needed for their jobs. But Braverman (1974) claimed with the deskilling theory that the skill content of most jobs was declining in spite of the increase in the educational attainment. In the 1980s and 1990s, more sociologists believed that high-skill jobs were relatively increasing in technology and sectoral shifts [(Attewell 1990, Form 1987, Wilson (1987, 1997)] and argued that this would create an increasing mismatch between the skills of minority workers and employer's workforce requirements. Katz and Murphy (1992) concluded that the growth in the demand for skill had overrun the supply more generally. The skills glut turned rapidly into a severe shortage. The US Department of Labor (1991) sought to clarify the skills needed for the work-ers and authorize new programs to set national occupational skill standards and strengthen the connection between education and workplace. Keep and Mayhew (1996), Krahn and Lowe (1998), and Payne (1999) asserted that Britain and Canada also engaged in an identical debate with similar urgency and anguish. Deep-rooted skills deficiencies were a major social problem and a principal barrier to social and economic improvement. Japanese econ-omy, the once-fearsome, seemed to be in the doldrums for nearly a dozen years, due to US's economic dominance in the 1990s. Moss and Tilly (2001) identified various attitudinal or demeanour issues, related problems with work motivation such as low effort and sense of responsibility as major problems. National Skill Development Corporation (NSDC) (NASSCOM, 2014) came out with its report that 500 million professionals would need to be skilled by 2022 to make them employable according to the National Skills Mission. The Economist Intelligence Unit (2015) reported that Indian companies were frustrated by graduates who were not workplace ready in servicing foreign clients. Chenoy et al. (2019) referred that under-skilled or semi-skilled workforce would pull down the global competitiveness of India as far as the productivity growth in manufacturing sectors was concerned. It is evident that the skill gap impedes the employability of the graduates and in the long run affects the economic productivity of the nation.

Studies on Communication Skills

Sivanganam's (2002) study on workplace communication skills for Malaysian graduates has highlighted the importance given to communication skills by the industry and the handicap of Malaysian engineering graduates due to their skills shortage. Mallett-Hamer's (2005) research on communication within the workplace has discovered the communication gap between the supervisors and the customer service representatives within the organization. The National Workforce Literacy Project (2010), in its report on workplace literacy and numeracy skills from employers' perspectives in Australia, has disclosed the assertion of 75% of employers that the businesses are affected by low levels of literacy in large enterprises and small and medium enterprises (SMEs). Some employers have reported that the engineers possessing high level of technical skills are poor communicators within workplace settings. Blom and Saeki (2011), in their study on the skill set of newly graduated engineers in India, have acclaimed communication in English to be the most demanded skills by the employers, as they are less satisfied with the skill set of the Indian engineers. Cai Yuzhuo et al. (2011–2012) have conducted a survey on employability of international graduates in Finland. The survey has highlighted that the graduates from India, China, and Russia have skills shortage and so the Finnish SMEs have found it difficult to induct them. The Finnish employers have opined that the technical students need more training in soft skills.

Singh et al. (2013) have carried out a research on the difference in perceptions of employers and instructors on the importance of employability skills in Malaysia. The study has addressed the mismatch between the skills that instructors perceive as important and those employers perceive to be important for employment. Both the employers and the instructors have selected communication skills as the most important generic skill required for the fresh graduates for their successful employment. Rahmat et al. (2012) have stated in their research on employment and graduates skill that the phenomenon of unemployment due to skills gap has occurred in many developing countries like India, Indonesia, the Philippines, and Thailand. Pandey and Pandey (2014), in their research paper "Better English for Better Employment Opportunities", have explored the surveys endorsing the skills gap in the Indian industries. The surveys have shown that only 10% of the graduates are employable out of several lakhs of students graduating engineering courses every year. The candidates are unsuitable because they lag in the soft skills which are essential for employability. Parvaiz (2014), in his thesis

"Skills Expectation-Performance Gap: A Study of Pakistan's Accounting Education", has identified oral communication and English language to be ineffective among various other 24 skills in students as expected by employers for employment purposes within the professional accounting institutes of Pakistan.

Clement and Murugavel (2015), in their case study on English for employability for engineering students in India, have stated the necessity of needs assessment in English for specific purposes (ESP) to create a meaningful learning environment and understand the needs and expectations of the learners that in turn enhances their employability skills. Karzunina et al. (2017) have highlighted the Global Skills Gap report that recommends liaising with employers to create work experience opportunities for students, facilitating closer correspondence between the skills universities teach and the skills businesses want. Deal (2017) has laid emphasis on the effectiveness of generic skills and technical education for the American manufacturing workforce by upskilling the skills to bridge the skills gap. Adrian (2017) has conducted a study on employers' expectations and students' perceptions regarding skills and has determined that the skills gap is the result of lacunae in oral communication skills among the new hires in management.

NASSCOM (2014) reports that in India, only 20% of engineering students are employable while the remaining 80% lack in communication skills. NSDC states that there will be a large man-power requirement in the engineering industries in India due to skill gap that is estimated to be 83.41% by 2017. The low level of ELP lessens the scope and prospects of employment. The graduates with low ELP might find a job but receive less effect on their immediate and desired work outcomes thus increasing the risk of career stagnation. The low ELP reduces the chance of being recruited and so a gap is created between industrial requirement and institute contribution. This necessitates an investigation into the level of English language competence possessed by the employees to understand their lacunae, and need that the present study has addressed the disadvantages of limited ELP of employees affecting the productivity and success of an organization in the long run and consequently has recommended certain ways to improve the communication skills of the students of engineering in educational institutions through workplace-based learning. The chapter tries to find out whether low proficiency in English degrades the reputation of the individual and the organization, the role of communication skills in English in improving the productivity of the organization, and to explore the skills gap existing between the employees' skill and the managerial requirement.

Research Design

A survey has been conducted between 2015 and 2016, within a period of nearly eight months, administering questionnaires to employees and employers of large, medium, small, and micro enterprises comprising 31 engineering industries dealing primarily with heavy boilers, fabrication, equipment, etc., nationwide. The data has been collected from 192 employee population comprising 63 Bachelors of Engineering (BEs) and 129 diploma holders in engineering discipline, working in various engineering enterprises in the district of Tiruchirappalli, Tamil Nadu, India. The employees' questionnaire consists of 35 questions with three parts. Part I deals with the social background of the employees such as their personal details, the place of their study period, their present designation in the workplace, and their previous working experience. Part II enquires the medium of instruction at the educational level, years of their study, English marks during their study, their educational background, their listening and reading habits, and the frequency of their English usage in schools and colleges. Part III questions their workplace management details such as their use of speaking and writing skills within and outside the workplace, their preparation for the skills improvement, and the mode or means of making up their deficiency.

The employers' questionnaire, distributed to 86 employers, deals with the requirement of the communication skills in English for a technical employee, employees' proficiency level in Listening, Speaking, Reading and Writing (LSRW) skills and the impact of the low ELP on the productivity of the organization. The statistical analysis of the data related to the mismatch between the employees' workplace performance and the managerial requirements substantiates the existing skills gap in the engineering enterprises in Tiruchirappalli district, Tamil Nadu. Thus, the following content emphasizes the prevailing skills gap highlighting how low ELP affects the reputation and the productivity of an organization, factors leading to skills gap, employers' views on skills gap, and remedial measures to bridge the skills gap with statistical evidences.

Low ELP Degrading the Reputation of the Organization

Employers' perceive that low ELP of the employees acts as the degrading factor to the reputation of the individual and the organization as highlighted in Table 3.1. As the results reveal, a substantial number of the employers have emphasized that the low ELP of the employees will affect the reputation of the individual and the organization in the outlook of the customers. Further, it indicates that the employees rarely interact in English and they are in need of an interpreter when they are unable to understand the instructions and directions given by the middle and the top managements. The employers find it strenuous to conduct meetings in English due to the poor communication skills of their employees, as is evident from Table 3.1. Hence, the low ELP of the employees results in the skills gap in the organizations.

It is evident from Table 3.1 that the customers weigh an organization, based on the ELP of the employees next to their technical knowledge. The reputation and the productivity are affected when miscommunication and

Table 3.1 Employers'Views on Low ELP Affecting Reputation of the Organization

Variables	Always	Sometimes	Rarely	Never
Low ELP affects reputation	67	18	1	-
Conduct of meetings in English	9	75	-	2
Employees' interaction in English	-	3	81	2
Need for interpreter	76	5	2	3
Employees' ELP	**Good**	**Average**	**Poor**	**Very Poor**
	3	81	2	-
Employees' writing proficiency	2	82	2	-

misconception occur between the employees of the company and its customers. Hence, the employers have attested low ELP as a degrading element in an organization. The inferences emphasize the employees' low proficiency level of English language and the prevailing skills gap that prove to be a menace to the reputation of the individual and the organization.

Significance of Communication Skills

Employers of today value skills such as effective communication, team work, leadership skills, problem-solving, and creativity on par with technical skills. Though there is a huge demand for engineers with good communication skills, the supply of graduates from engineering colleges and polytechnics is meagre to match this expected demand.

Table 3.2 explicates the role of communication skills in improving the productivity of an organization. When the employers have been asked about the contribution of communication skills in advancing the career of employees, majority of the employers have emphasized the major contribution of communication skills in advancing one's career. When the importance and requirement of communication skills at the workplace have been asked to the employers, they have presented their opinion towards all the variables positively, as apparent in Table 3.2. The employers have opted 'Always' as their first priority for the requirement of communication skills in the career of an engineer. They have substantiated the employees' recurrent communication in English with foreign or state customers in their organization. All the employers have admitted that the communication skill is a value addition for a technical employee. The employees have presented their opinion on the purpose, kind, and form of workplace communication. They have ranked 'seeking clarifications from customers' as their chief purpose for using English. They have indicated that they involve more writing than speaking in English. They have expressed that they often use Email for workplace communication. The mean values of the variables in Table 3.2 highlight the priority on the requirement of communication skills in an organization. Hence, it implies that the communication skills of an employee

Table 3.2 The Impact of Communication Skills on the Productivity of an Organization

Variables	Always (%)	Sometimes (%)	Rarely (%)	Never (%)	M	SD
Career advancement	94.4	5.6	-	-	3.94	0.24
Requirement in an organization	66.7	33.3	-	-	3.67	0.49
Frequency of communicating in English	44.4	22.2	33.3	-	3.11	0.90
Value addition	Yes 100.0		No -		2.00	0.00
Employees' opinions						
Purpose of English use at workplace	To converse with co-workers 23.4	To understand directions from managers 29.2	To understand the instructions from CEOs 13.0	To seek clarifications from customers 34.4	2.58	1.19
Kind of workplace communication	Reporting supervisors in writing 42.7		Speaking to co-workers 16.1	Both 41.1	1.98	0.92
Form of workplace communication	Telephone 10.4	Email 77.1	Letters 9.4	Reports 3.1	2.95	0.57

help him progress in his career as it enables him to converse well with the foreign customers in English thus proving himself to be an instigator of the productivity of the organization. An organization faces pitfalls due to the low proficiency of the employees. The employers find it strenuous to conduct meetings in English, as the employees fail to understand the intended message passed on in the second language. The employees hesitate to communicate in English within the workplace due to their paucity in communication skills. The employers are in a deplorable condition to correct the errors committed by their employees in the written communications in order to save the reputation of the organization among the customers. All these drawbacks stand as causes for the skills gap in the organizations.

Aspects of Communication that Lead to Productivity

Internal communication, clientele conversation, and customer satisfaction play major roles in improving the productivity of an organization. Among these options, both the employers and the employees have endorsed internal communication to be the vital aspect of communication at the workplace as exhibited in Table 3.3. Effective communication among the top, middle managements and the employees will enhance proper understanding of messages conveyed and enable appropriate execution of the instructions. This bridges the skills gap leading to a remarkable productivity in the organization.

Table 3.3 highlights the perspectives of the employers and the employees on the contribution of communication skills and the aspects of communication to the improvement of the productivity in the organization. Most of the employers have preferred the range of 70% to 90% of communication skills as the requirement to improve the productivity. A majority of the employers have admitted the fact that the communication skills enhance the

Table 3.3 Employers' Views on the Aspects of Communication That Lead to Productivity

Variables	Less than 50%	50%–70%	70%–90%	90%–100%
Productivity and quality service	2	15	64	5
Client relationship, rapport, and trust	**Always** 80	**Sometimes** 6	**Rarely** -	**Never** -
Client needs and expectations	84	2	-	-
Aspects of communication	**Internal communication**	**Clientele conversation**		**Customer satisfaction**
Employers	69	15		2
Employees	84	29		78

client relationship, rapport, and trust. It also helps the employees to understand the client needs and expectations. Besides, a substantial number of the employers and the employees have opted 'Internal communication' to be the topmost communication aspect to augment an organization's productivity. So, it is inferred that if the employees initiate to interact in English with the managers in their workplaces, that in turn would instill confidence and enhance their communication skill to communicate with their CEOs and the clients. This eventually results in the customer satisfaction thus taking the orders and the productivity to the high level.

Skills Gap Prevailing in the Enterprises in Trichy District

Table 3.4 exhibits the skills gap prevailing in the enterprises in Tiruchirappalli district and the reasons for the skills gap at the workplace with reference to the views of the employers. Though there are many factors that lead to skills gap at the workplace, the prevailing skills gap in the organizations is due to the mismatch between the expectations of the employers and the performance of the employees. When the employers have been asked to select the predominant reason for the skills gap prevailing in the organizations, nearly 95% of the employers have remarked the employees' low proficiency in the communication skills as the primordial reason and the employees' educational system as the secondary reason. It is indicated in Table 3.4 that the percentage of skills gap in the manufacturing sectors is considerably high. Hence, it is inferred that the prevalent reason for the prevailing skills gap is due to the employees' inadequacy in communication skills and

Table 3.4 Employers' Views on Skills Gap and the Reasons for It

Variable	Category	Frequency	Per cent
Reasons for the skills gap	Prevailing educational system not on par with industry requirement	28	32.6
	Satisfaction of customers with less quality goods	-	-
	Lack of commitment from the employees' side	I	1.2
	Low proficiency of communication skills of the employees	57	66.3

	Yes		**No**	
Skills gap exists between the employees' skill and the managerial requirement	**Frequency**	**Per cent**	**Frequency**	**Per cent**
	82	95.3	4	4.7

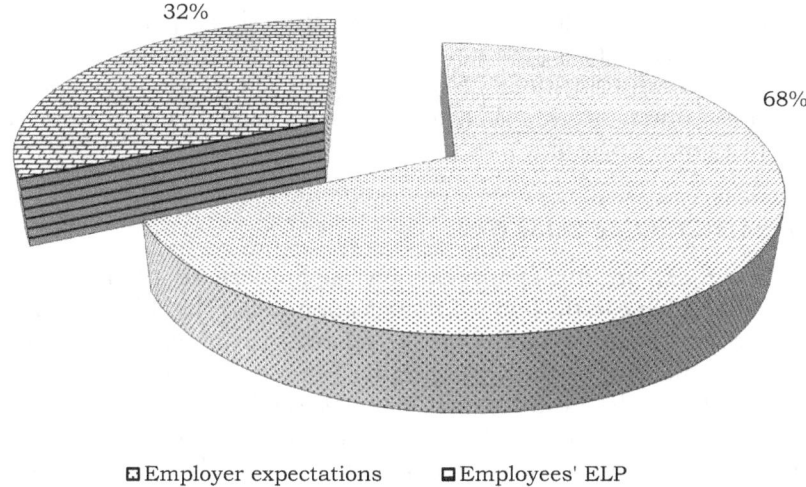

32%

68%

◻ Employer expectations   ◻ Employees' ELP

Figure 3.1 Expectations of Employers vs ELP of Employees.
Source: Authors.

the low proficiency in English. The employees of an organization need to possess requisite communication skills along with technical knowledge to contribute better in industrial productivity. The prevailing skills gap can be eliminated by facilitating a skill-based curriculum to suit the industrial requirements in the world of work.

Figure 3.1 shows the skills gap prevailing in the organization due to the difference between the expectations of the employers on communication skills in English and the lack of ELP of the employees.

Lack of Communication Skills and Skills Gap

Table 3.5 shows the selection of employees in the enterprises based on their communication skills. The table highlights the employers' views on the evaluation of the interviewees' communication skills during the recruitment

Table 3.5 Employers' Evaluation of Communication Skills during the Recruitment Process

Look out for good communication skills during selection process	Always	Sometimes	Rarely	Never
	58	25	3	-

process. A substantial number of employers have asserted that they 'always' select candidates with requisite communication skills apart from their technical knowledge during the interview.

It is inferred from Table 3.5 that it is pertinent to select candidates with requisite communication skills to avoid skills gap and depreciation in productivity of the organizations in the long run.

Factors Creating Skills Gap

Table 3.6 explores the factors on how the productivity of an organization gets affected due to the poor communication skills. The chi-squared test has been computed to show the factors related to communication skills that affect productivity of the organizations over a period of time.

In Table 3.6, the P value greater than 0.05 signifies that there is no association between the employers' perspectives on the enhancement of productivity and the employees' views on English use at workplace, say understanding the instructions, speaking, and reporting in English and difficulty in drafting emails. On the other hand, there is a significant association between the employers' perspectives on increasing the productivity of the organization and the employees' opinion on improving their communication skills. It is inferred that if the employees improve their communication skills, the productivity of the organization will also improve. Proportionately, when the employees fail to use English effectively at the workplace, the rapport with the clients and customers gets affected. The employees' inability to understand instructions in English given by the managers and CEOs, and execute them efficiently by speaking and reporting to their supervisors in English will create lapse in the conveying of messages. Harmer (2001, p. 272) has stated, "the ability to speak fluently presupposes not only knowledge of language features but also the ability to process information and language 'on a spot'". Such skills gap hinders the improvement of productivity in the organization.

Table 3.6 Factors Leading to Skills Gap by Affecting Productivity

Variables	Chi Square	Probability Value	Significant Value	Inference
Use of English	2.107	0.910	P >0.05	**NS**
Understanding of instructions	11.887	0.521	P > 0.05	**NS**
Speaking and reporting in English	2.613	0.978	P > 0.05	**NS**
Difficulty in drafting emails	5.818	0.758	P > 0.05	**NS**
Improving communication skills	22.073	0.037	P < 0.05	**Significant**

Remedial Measures to Improve Communication Skills

Table 3.7 explicates the remedial measures opted in improving the communication skills of the employees. The employers and the employees have been asked to choose the remedial measures to make up the deficiency in the communication skills. About 46.5% of the employers have preferred the incumbents to attend spoken grammar classes to improve their communication skills. About 74.4% of the employees, on the other hand, have expressed that they can improve their communication skills by learning books by themselves.

It is inferred from Table 3.7 that the skills gap persists in the organizations as is evident from the perspectives of employers and employees resorting to remedial measures. Apart from the suggested remedial measures, a paradigm shift in the curriculum and policy-making decisions in the educational system has to be implemented so as to upgrade the employability skills of technical students. It is also inferred that the incumbents stammer to speak and write sentences in English without grammatical errors due to their poor communication skills. Students of engineering disciplines learn English only to score in the examinations by cramming words and grammar. They do not consider the activity of learning English to be more important than logical and mathematical activities. Hence, they are likely to end up with communication failures when they encounter certain tasks that involve communication in English at the workplace.

Findings and Discussion

The present study explores the skills gap prevailing in the workplace of the engineering sectors and how the reputation of the individual and the organization gets affected due to the dearth in employees' skill set. Table 3.1 shows the employers' perspectives that low proficiency level of employees sometimes degrades the companies' and employees' reputation. The clients and customers are usually contented with the organizations when they are provided with the appropriate and timely response pertaining to the product they expect during delivery. Technical engineers need communication skills when engineering projects are planned and executed across the national and cultural borders. Deficiency in requisite skills set would result

Table 3.7 Employers' Opinion on the Remedial Measures

Remedial measures for the deficiency in English language skills	Employers (%)	Employees (%)
Learning from self-study books in English	31.3	74.4
Attending grammar classes	46.5	0.08
Attending spoken English classes	22	16.6

in barriers for the engineers' personal and professional development. The frequent discussions with openness and transparency with the team members would help to communicate in English without any hesitation and to maintain the levels of trust within and outside the company.

Table 3.1 elicits the employers' views on the low ELP that affects the reputation of the organization. The employers are in a despicable condition to accept that their employees need the help of an interpreter to explain the messages which they do not understand. The hesitation and reluctance of the employees to interact in English within the workplace due to their inability is pitiable. Table 3.2 highlights the significance of communication skills in improving the productivity of organization. The results emphasize the requirement of communication skills at the workplace for career advancement and communicating with state and foreign customers. The employers have expressed their embarrassing situation of conducting scanty meetings in English due to the poor communication skills of their employees.

The upward communication should be effective between middle management comprising managers and top management headed by senior executives. The communication between managers and employees should be clear and effective in such a way that the key practical roles in the company are properly carried out. It is in this downward communication that the middle managers need to address cold-call salesmen in the sales department to a machine operator at a production plant. They are expected to cater their teams with company information updating on business priorities, business initiatives, and organizational strategy. Ineffective communication in this regard will lead to disruption across the workforce, ultimately affecting morale, efficiency, and productivity. The level of requirement of communication skills at workplace is tabulated by frequency percentage. The employees' priority of the purpose, kind, and form of workplace communication is indicated through in Table 3.2. It also shows that the Communication skills as a value addition play a pivotal role in improving the productivity and quality service in an organization.

Table 3.3 has been computed to show the communication aspect that improves productivity. Both the employers and the employees have unanimously approved internal communication to be the topmost aspect of communication in improving an organization's productivity. Table 3.4 enumerates the employers' views on skills gap and the reasons for its prevalence in the workplace. The employers have endorsed employees' low proficiency of communication skills to be the prime reason for the skills gap acceding to the inadequacy of the current educational system in meeting out the industrial requirements.

Figure 3.1 vividly epitomizes the existing skills gap through the expectations of the employers and the level of ELP of the employees. At the entry level, the importance of selecting candidates with good communication skills in English into the organizations is emphasized in Table 3.5. Table 3.6 elicits how the lack of communication skills at the workplace affects the productivity accelerating skills gap. It strongly emphasizes the skills mismatch embedded in the organizations irrespective of the large or medium and small manufacturing enterprises (MSMEs). Table 3.7 elucidates the remedial measures suggested by the employers to eliminate the prevailing skills gap in the enterprises in Tiruchirappalli district. The comparative analysis of the opinions of the employers and the employees highlights a substantial disparity between the opinions of the groups. It is evident that the mismatch between the performance of the employees and the expectations of the employers affect the reputation of the organization in general and the CEOs in particular. This reinstates the fact that lack of communication skills of the employees has created the skills gap in the enterprises, thereby affecting the productivity of the organizations.

Recommendations

Smith et al. (2007) have opined that the students must be provided with transferrable skills that can be taken into the workplace. Four sources of weakness that mitigate the career prospects of the engineers in communication skills education are identified by Roulston and Black (1992) as,

- Students' attitudes to communication
- Insufficient course content
- Deficient and inappropriate teaching methods
- Lack of opportunity to practice communication skills

In recent times, business enterprises realize the importance of collaborating with educational institutions to bridge the skills gap in workplace situations. Patricia Claghorn, Dean of Continuing Education and Institutional Advancement for Gloucester County College (GCC) in Sewell, New Jersey (2012) [as cited in Bridging the Skills Gap (2012) p. 11], has affirmed that colleges need business enterprises to serve as the working advisors so that the curriculum can have relevance and value for their organizational goals and employees' knowledge and skills. This collaboration will create focus on the new business training for the colleges on how higher education and employers can work together to increase the educational level of the employees.

The World Economic Forum (2015) has stated that organizations can attract and absorb knowledgeable and skilled persons into their workforce by partnering with all levels of government and accessing training resources. The educators have to ensure liaising with employers at local, regional, and global level and create work experience opportunities for as many students as possible. They have to engage fervently to enhance the standard of the skills taught by the institutions and the skills expected by the employers.

Educational institutions are supposed to teach soft skills and supply as per the demand of the workplace environment. Oxford Economics (2012) has claimed that the new business structures give rise to an additional demand for a wide array of skills from technical knowledge to soft skills such as interpersonal communication, agile thinking, etc. Kember and Leung (2005) have aspired that they must nurture the appropriate competencies and consider the best to ensure that the skills are developed. They have to seek ways to integrate employability-focused tenets into curricula. Goleman (1996) has suggested that the competencies such as communication skills, teamwork-related skills, and critical thinking abilities that are valued by the employers could be designed by the educational institutions to be taught in addition to the core subject skills.

According to Bath et al. (2004), soft skills should be an important element in the undergraduate programmes as they are necessary in any field of work. The India Study Channel (2013) [as is cited in Ganesh, 2013] has stated that communication as a soft skill does not involve what one communicates and how he communicates. It makes the communication differ from one situation to the other and one receiver to the other by pruning the language, body language, expression, tone of speech, ability to listen and understand, ability to find solutions, attitude, manners, concern about audience, and other related skills. It is the responsibility of higher educationalists to incorporate this as a part of their teaching and learning process. Rainsbury et al. (2002) have emphasized on the insufficient importance placed on the development of soft skills by higher education institutions. Hind et al. [as cited in Kavita and Sharma, 2011, p. 38] have envisaged the need for employability skills to be embedded throughout the curriculum instead of confining to a single module. Vignali and Hodgson (2007) have stressed the responsibility of the higher education to provide its graduates with the necessary skills to operate them professionally within the work environment. Tamil Nadu Skill Development Corporation (TNSDC, 2019) has to initiate the upskilling of state's youth by developing a custom pack of the interventions, encompassing 21st century employment skills and soft skills. The graduates' exposure to work environments through industry visits, internships, and apprenticeships must be strengthened.

It is imperative that the educational institutions in India must take initiatives to ensure the employability of the graduates passing out every year. The onus is on the education system to introduce and develop curriculum that imparts skills relevant to the industry. The National Education Policy (2019) by the Indian Government has implemented professional education to be the integral part of the higher education system. Curriculum focusing on LSRW skills from grades 4 and 5 to be taught sequentially is introduced. English language is given due importance along with the study of other regional languages. PSG College of Technology in Peelamedu, Coimbatore, is one of the engineering colleges in Tamil Nadu to impart industrial training for students in the fifth year of B.Tech/B.E.-Sandwich course for all the disciplines while the regular courses run for four years as in other institutions. Aspiring Minds–AMCAT (2014) convenes a four-year program that bridges the skills gap by helping both institutions and their students in evaluating their job readiness by realizing their strengths and weaknesses.

This programme equips the students to know their efficiency on the type of jobs when they reach the final year of the college and with the right skill set at the time of interview. The Indian Government has launched National Employability Enhancement Mission (NEEM) in association with the AICTE (Berry, 2018). It gives 'on the job training' program under four semesters to the students; the first part being the classroom sessions and the second part being the real ground of industry. The company under NEEM scheme trains the students with latest tools, machines, and technology and enhances the employability of the students which the institutes lack. Bureau of Industrial Consultancy Services (BICS) of Jawaharlal Nehru Technological University (JNTU), Hyderabad, functions as university's interface with industry. National Institute of Technology, Tiruchirappalli (NIT-T), Tamil Nadu, gives hands-on training to students with the Siemens Manufacturing Training unit within its campus. The study suggests to impart industrial training to the students by the educational institutions in order to bridge the gap surfacing between academia and industry.

Conclusion

Skills gap affects companies' ability to grow, innovate, deliver products and services on time, and meet quality standards. Bridging skill gaps in the formal or informal sector would improve employment, enterprise creation, and productivity. This chapter has attested how lack of employees' ELP leads to productivity lapse, encompassing the consistent emphasis of the employers on the need for communication skills and the factors of skills gap through the statistical data. It has focused on academics being combined with practical training to meet the shortage of demand supply. It stresses on the dire need of revamping the curriculum and bring in new teaching methods and technology. The chapter has vehemently emphasized the requirement of communication skills for the engineers and the need to bridge the skills gap persisting in the manufacturing enterprises. The computed data reiterates the role of communication skills in enhancing the productivity of the organizations. The industry-academia interface is thus, recommended for the skilled workforce to work effectively in competitive workplace environment, resolve problems confidently, comprehend stakeholders' needs, and cope with the rapidly changing business environment.

References

Adrian, M. (2017). Determining the skills gap for new hires in management: Student perceptions Vs employer expectations. *International Journal for Innovation Education and Research, 5,* 139–147.
Aring, M. (2012). *Skills gaps throughout the world: An analysis for UNESCO global monitoring report 2012: Report on skills gap.* United Nations Educational, Scientific and Cultural Organization. Available at: http://unesdoc.unesco.org/images/0021/002178/217874c.pdf [Accessed 10th November 2018].

Aspiring Minds. (2019). *National employability report-engineers-annual report 2019*. NER_Engineer_2019_V5.pdf

Aspiring Minds–AMCAT. (2014). *Why are Indian engineers unemployable?* Retrieved from http://www.aspiringminds.in/docs/national_employabilityReport_engineers_annual_report_2014.pdf.

Attewell P. (1990). What is skill? *Work and Occupations, 17,* 422–448.

Bath, D., Smith, C., Stein. S., & Swann, R. (2004). Beyond mapping and embedding graduate attributes: Bringing together quality assurance and action learning to create a validated and living curriculum. *Higher Education Research and Development, 23*(3), 313–328.

Berg I. (1971). *Education and jobs: The great training robbery*. Boston: Beacon Press.

Berry, R. (2018, April 28). CLR skills launches skill development program in India under NEEM scheme.

Blom, A., & Saeki, J. (2011). Employability and skill set of newly graduated engineers in India (English). *Policy research working paper No. WPS 5640.*South Asia. Retrieved from http://documents.worldbank.org/curated/en/45588146826 7873963/Employability-and-skill-set-of-newly-graduated-engineers-in-India

Braverman, H. (1974). *Labor and monopoly capital*. New York: Monthly Review Press.

Bridging the Skills Gap. (2012). *Help Wanted, Skills Lacking: Why the Mismatch in Today's Economy?* [online] Available at: https://www.nist.gov/sites/default/files/documents/mep/Bridging-the-Skills-Gap_2012.pdf [Accessed 8th May 2018].

Cai, Y., Shumilova, Y., & Pekkola, E. (2011–2012). *Employability of international graduates educated in Finnish higher education institutions*. VALOA-project Career Services, Finland: University of Helsinki.

Career Builders. (2017). *74% of employers say they plan to hire recent college graduates this year, according to annual career builder survey*. [online] Available at: http://press.careerbuilder.com/2017-04-27-74-Percent-of-Employers-Say-They-Plan-to-Hire-Recent-College-Graduates-This-Year-According-to-Annual-Career-Builder-Survey. [Accessed 10th May 2018].

Chenoy, D., Ghosh, S. M., & Shukla, S. K. (2019). Skill development for accelerating the manufacturing sector: The role of 'new-age' skills for 'Make in India', *International Journal of Training Research, 17*(1), 112–130. doi: 10.1080/14480220.2019.1639294

Clement, A., & Murugavel, T. (2015). English for employability: A case study of the English language training needs analysis for engineering students in India. *English Language Teaching, 8,* 116.

Collins R. (1979). *The credential society: An historical sociology of education and stratification*. New York: Academic Press.

CompTIA. (2012). *State of the IT Skills Gap*. [online] Available at: https://www.wired.com/wp-content/uploads/blogs/wiredenterprise/wp-content/uploads/2012/03/Report_-_CompTIA_IT_Skills_Gap_study_-_Full_Report.sflb_.pdf. [Accessed 8th May 2018].

Deal, R. M. (2017). *The skills gap in U.S. manufacturing: The effectiveness of technical education on the incumbent workforce*. [Unpublished M.Sc. Dissertation]. Western Kentucky University, Kentucky.

Economist Intelligence Unit Ltd. (2015, January). *Skills needed: Addressing South Asia's deficit of technical and soft skills: Analysing the gap in Afghanistan, Bangladesh, India, Nepal, Pakistan and Sri Lanka*. Retrieved from https://www.

britishcouncil.org.bd/sites/default/files/skills_needed_addressing_south_asias_ deficit_of_techincal_and_soft_skil.pdf

Form, W. (1987). On the degradation of skills. *Annual Review of Sociology, 13,* 29–47.

Ganesh, S. (2013). Importance and necessity of soft skills in professional life. *India Study Channel.* Available at: http://www.indiastudychannel.com/ resources/159706-Importance-and-necessity-of-soft-skills-in-professional-life. aspx. [Accessed 7th September 2018].

Goleman, D. (1996). *Emotional intelligence.* London: Bloomsbury.

Harmer, J. (2001). *The practice of English language teaching.* Edinburgh, Harlow: Pearson Education Limited.

India Skills Report. (2017). *Reaching Over 5,60,000 Students Across 29 States, 7 Union Territories and 125+ Employers.* Available at: https://www.aspiring- minds.com/sites/default/files/National%20Employability%20Report%20-%20 Engineers%20Annual%20Report%202016.pdf

Karzunina, D., West, J., Moran, J. & Philippou, G. (2017). *The global skills gap: Student misperceptions and institutional solutions.* Philadephia: Wharton University of Pennsylvania. Retrieved from https://www.reimagine-educa- tion.com/wp-content/uploads/2018/01/RE_White-Paper_Global-Skills-Gap- Employability.pdf

Katz L. F., & Murphy, K. M. (1992). Changes in relative wages, 1963–1987: Supply and demand factors. *Quarterly Journal of Economics, 107,* 35–78.

Kavita, K. M., & Sharma, P. (2011). Gap analysis and skills provided in hotel man- agement education with respect to skills required in the hospitality industry: The Indian scenario. *International Journal of Hospitality and Tourism Systems, 4*(1), 31–51.

Keep, E., & Mayhew, K. (1996). Evaluating the assumptions that underlie training policy. In A. Booth & D. Snower (Eds.) *Acquiring skills, market failures, their symptoms and policy responses* (pp. 305–334). Cambridge: Cambridge University Press.

Kember, D., & Leung, D. (2005). The influence of the teaching and learning environ- ment on the development of generic capabilities needed for a knowledge-based society. *Learning Environments Research, 8*(3), 245–266.

Krahn, H., & Lowe, G. S. (1998). *Literacy utilization in Canadian workplaces.* Ottawa: Minister of Industry. Catalogue no. 89-552-MIE, 4.

Mallett-Hamer, B. (2005). *Communication within the workplace.* [Unpublished Master's Dissertation]. University of Wisconsin-Stout, Menomonie.

Moss, P., & Tilly C. (2001). *stories employers tell: Race, skill, and hiring in America,* New York: Russell.

NACE. (2016). *Job Outlook 2016: The Attributes Employers Want to See on New College Graduates' Resumes.* Available at: http://www.naceweb.org/career-devel- opment/trends-and-predictions/job-outlook-2016-attributes-employers-want-to- see-on-new-college-graduates-resumes/ [Accessed 9th May 2018].

NASSCOM. (2014). *Analysis of Talent Supply and Demand: Employment Requirements and Skill Gaps in the Indian IT-BPM Industry (NSDC).* Retrieved from https://s3-ap-southeast1.amazonaws.com/pursuiteproduction/media/ Reports/Analysis%20Report%20Final%2017.02.2014.pdf

National Workforce Literacy Project. (2010). *Report on employers' views on work- place literacy and numeracy skills.* Australia: Australian Industry Group.

Oxford Economics. (2012). *Global Talent 2021: How the New Geography of Talent will Transform Human Resource Strategies*. Available at: https://www.oxfordeconomics.com/Media/Default/Thought%20Leadership/global-talent-2021.pdf

Pandey, M., & Pandey, P. (2014). Better English for better employment opportunities. *International Journal of Multidisciplinary Approach and Studies, 1*, 93–100.

Parvaiz, G. S. (2014). *Skills expectation-performance gap: A study of Pakistan's accounting education*. [Unpublished Doctoral Dissertation]. London: Brunel University.

Payne, J. (1999). *All things to all people: Changing perceptions of 'skill' among Britain's policy makers since the 1950s and their implications*. SKOPE Research Papers. University of Warwick, SKOPE. ESRC Centre on Skills, Knowledge and Organisational Performance (SKOPE). Retrieved from http://www.skope.ox.ac.uk/?person=all-things-to-all-people-changing-perceptions-of-skill-among-britains-policy-makers-since-the-1950s-and-their-implications

Rahmat, M., Ahmad, K., Idris, S., & Zainal, N. F. A. (2012). Relationship between employability and graduates' skill. *Procedia - Social and Behavioral Sciences, 59*(4), 591–597.

Rainsbury, E., Hodges, D., Burchell, N., & Lay, M. (2002). Ranking workplace competencies: Student and graduate perceptions. *Asia-Pacific Journal of Cooperative Education, 3*(2), 8–18.

Roulston J. D., & Black R. W. (1992). Educating engineers: What's happening to communication? *4th annual convention and conference Australasian association for engineering education* (pp. 190–193). Brisbane, Australia.

Singh, P., Thambusamy, R., Ramly, A., Abdullah, I. H., & Mahumad, Z. (2013). Perception differential between employers and instructors on the importance of employability skills. *Procedia-Social and Behavioral Sciences, 90*, 616–625.

Sivanganam, L. (2002) *Communication skills: What do employers expect? (Workplace Communication Skills for Engineering Graduates*. [Unpublished Master thesis]. Universiti Tun Hussein Onn Malaysia. Retrieved from http://eprints.uthm.edu.my/id/eprint/909/

Smith, K., Clegg, S., Lawrence, E., & Todd, M. J. (2007). The challenges of reflection: Students learning from work placements. *Innovation Education and Teaching International, 44*(2), 131–141.

TNSDC. (2019). Skilling for the future: Skill gap assessment and action plan for Tamil Nadu, November 2019. Available at: Microsoft Power Point—TNSDC State Report—Cover Page LAST. pptx.

The Conference Board. (2008). *Fit for Purpose-are China's Graduates Ready to Work?* [online]. Available at: http://unesdoc.unesco.org/images/0021/002178/217874e.pdf. [Accessed 8th May 2018].

US Department of Labor. (1991). *What Work Requires of Schools: A SCANS report for America 2000*. Retrieved from http://www.gsn.org/web/_shared/SCANS2000.pdf

Vignali, G., & Hodgson, I. (2007). Real world learning – Enhanced employability. *EUROCHRIE conference*. Leeds, UK. Available at: http://www.celt.mmu.ac.uk/ltia/issue18/vignali.php

Wilson, W. J. (1987). *The truly disadvantaged*. Chicago: University of Chicago Press.

Wilson, W. J. (1997). *When work disappears: The world of the new urban poor*. New York: Vintage.

World Economic Forum. (2015). *Bridging the skills and innovation gap to boost productivity in Latin America: The competitiveness lab – A World Economic Forum initiative*.

Chapter 4

Task-Based Language Teaching to Foster Communication Skills

N. Chitra

Modern technological era demands more efficient engineers who are capable of confronting problems of a professional scope with confidence. Communicative skills play a crucial role in recreating a true work environment as well as maintaining a good rapport in the society. For a professional student, it is essential to be proficient both in academic and interpersonal functions. The increasing degree of unemployability of engineering students in the present scenario brings students' substandard skills and inadequate competence to the fore. Hence, initiatives have been taken recently in the education system, to strengthen the skills of the students to meet the changing social and professional demands. Recognizing the learners' needs, the English syllabus has been redesigned with the focus on a competence-based approach. Learning a second language for communicative purposes requires knowledge about the structural and other aspects of linguistic forms as well as skills for using it appropriately. Learning efficiency of communication skill would be lost when it is taught as other subjects. As a skill, it could be practiced by creating 'meta learners' through 'meta teaching' making learners responsible for their own learning. The language teachers could make the students conscious of their learning process and hopeful of their success, insisting on their need for diligence to master the language. The task-based language learning (TBLL), in which language learning is accomplished by performing tasks, could be the appropriate approach to achieve the desired objectives. The present chapter is a study of experiential learning of language and communication through task-based language teaching (TBLT) methodology reviewing its fundamentals and the required roles of teachers and students, besides explaining its benefits and challenges along with findings and recommendations.

Task-Based Language Teaching

During 1990s, TBLT has become a powerful and advanced learning and teaching approach, advocating 'tasks' as its main pedagogical tools to structure language teaching. It has started receiving sustained attention from a number of theoreticians, practitioners, and researchers in language

DOI: 10.4324/9781003268529-6

education. A few theoreticians have believed TBLT to be the extension of communicative language teaching (CLT) in its strong form. Many have observed that TBLT has emerged in response to some constraints of the traditional PPP approach, denoted by the process of presentation, practice, and performance (Ellis, 2003).

TBLT emphasizes the idea that language learning is a developmental process enhancing communication and social interaction rather than a product internalized by practicing language items. It believes that learners master the target language more powerfully when being exposed to multiple, meaningful, and challenging task-based activities in a natural way. In the process of performing tasks, it provides rich opportunity for the learners to interact with peers, comprehend each other, and present their own meaning, which ultimately promotes their language knowledge and professional skills. In a nutshell, TBLT is an approach that "aims to develop learners' communicative competence by engaging them in meaning-focused communication through the performance of tasks." (Ellis & Shintani, 2014, p. 135). Communicative competence can be defined simply as "what a speaker needs to know to communicate appropriately within a particular language community" (Saville-Troike, 2010). It can be acquired through consistent use of the second language fulfilling both academic and interpersonal functions. Knowledge about traditional levels of language such as vocabulary, morphology, phonology, syntax, and discourse in accordance with listening, speaking, reading, and writing skills (LSRW) activities would enable learners to attain pragmatic competence through the practice of tasks.

Tasks in EFL Context

TBLT revolves around its core concept 'task' in relation to teaching, learning, and assessing. The term 'task' in the English as a Foreign Language (EFL) context has attracted many theoreticians to such an extent that they define it in various ways. Table 4.1 summarizes the key concepts of definitions of 'tasks' by different researchers.

Long (1985, p. 89) provides non-pedagogical definition for 'task' expounding it as 'a piece of work undertaken for oneself or for others freely or for some rewards… It is non-technical and non-linguistic'. Distinguished linguists like Breen (1987), Willis (1996), Skehan (1998), Ellis (2003), and Nunan (2005) propagate pedagogical tasks that are carried out in the classes focusing on real-world activities of everyday life. They believe 'task' to be more effective means of promoting Second Language Acquisition (SLA) in the class room in more natural, meaningful way. Breen (1987) considers 'task' as "a range of work plans which have the overall purposes of facilitating language learning from the simple and brief exercise type to more complex and lengthy activities." Littlejohn's (1998) insight of 'task' has insisted on the three aspects of process, participation, and content of task and has explicated that 'process' means what teachers and learners go

Table 4.1 Definitions of Task

Researchers	Key concepts
Long (1985)	What people do in everyday life, at work, at play, and in between?
Breen (1987)	A range of work plans for exercise and activities in language instruction.
Skehan (1998)	Meaning, task completion, the real world, and outcome are focused.
Willis (1996)	A classroom undertaking for a communicative purpose to achieve an outcome.
Ellis (2003)	A work plan that requires learners to process language pragmatically to achieve an outcome.
Nunan (2005)	A piece of classroom work to convey rather than to manipulate form.

through, 'content' is something that learners focus on and 'participation refers to whom learners work with in the process. Skehan (1998) propounded that tasks primarily focus on meaning and resemble real-life situations. Willis (1996) is also of the view that a 'task' is an activity "where the target language is used by the learner for a communicative purpose (goal) in order to achieve an outcome" (p. 23). Students could express their thoughts even if some of the language is inaccurate. Thus, many researchers have attempted to define the term 'task' emphasizing different linguistic aspects.

The definition of task has been improved after Ellis (2003) has merged numerous views in task-based research and pedagogy. Emphasizing both meaning and form of language, Ellis (2003) defines 'task' as "A work plan that requires learners to process language pragmatically in order to achieve an outcome... Like other language activities, a task can engage productive or receptive, and oral or written skills and also various cognitive processes" (p. 16). This definition proposed by Ellis (2003) embodies the six criteria of a 'task' which can be stated as follows:

- A task can be considered as a work plan.
- In a task, the main focus is on meaning.
- A task includes everyday processes of language use.
- A task can comprise any of the four language skills.
- A task involves cognitive processes.
- A task has a clearly defined communicative result.

Nunan (2005) defines 'task' as "a piece of class room work that involves learners in comprehending, manipulating, producing or interacting in the target language while their attention is focused on mobilizing their grammatical knowledge in order to express meaning rather than to manipulate form." For him, 'tasks' are classroom undertakings providing a context and are intended to result in pragmatic language use to promote L2 learning. On the whole, 'tasks' are goal oriented, meaningful activities designed to facilitate students' participation by exposing them to language production and reception.

Task-Based Approach in Second Language Teaching

Task-based approach in second language teaching has been first performed by Prabhu under the Communicational Teaching Project (CTP) (1979–1984) in Bangalore, India. As a reaction against the dominance of structural approach to a language, the CTP has strongly recommended that the structure of a language is best acquired unconsciously without the learner's attention being focused on the language. Prabhu (1987) believed "Grammar construction by the learner is an unconscious process which is best facilitated by bringing about in the learner a preoccupation with meaning, saying or doing." Learners might learn more efficiently when the focus of their minds is more on the task, rather than on the language they are using (Prabhu, 1987). Though CTP has combined task-based syllabus and communicational methodology, it did not insist on practicing real-life activities. The core functions and value of TBLT in constructing learner-centered classrooms and language learning contexts, providing learners the chance to communicate and interact, and enhancing learners' ability to deploy the target language and sort out communicative problems have been highly appreciated and recognized by researchers in the area of language teaching in the 1980s. Recent studies on TBLT highlight its three recurrent features: "TBLT is compatible with a learner-centered educational philosophy (Ellis, 2003; Nunan, 2005; Richards & Rodgers, 2001), it consists of particular components, such as goal, procedure, specific outcome (Murphy, 2003; Nunan, 2005; Skehan, 1998), and it advocates content-oriented meaningful activities rather than linguistic forms (Beglar & Hunt, 2002; Carless, 2002; Littlewood, 2004)".

Characteristics of TBLT

Swan (2005) emphasizes the following characteristics of TBLT after careful analysis of its core principles propagated by divergent theoreticians and practitioners:

1. Instructed language learning should mainly contain natural and naturalistic language use, and the activities are related to meaning rather than language.
2. Instruction should support learner-centeredness rather than teacher-centeredness.
3. As learning is promoted through natural way, target like accuracy is not aimed at.

4. Engagement is essential to promote internalization of formal linguistic elements.
5. Communicative tasks are especially suitable devices for such an approach.

For a better practice of TBLT, Nunan suggests eight principles to be followed:

1. **Scaffolding** through lessons and materials that provide support to the students.
2. **Task chains** in which exercises, activities, and tasks have provided relevance and coherence.
3. **Recycling** language to maximize opportunities for learning.
4. **Organic learning** in which language ability grows gradually.
5. **Active learning** through which learners actively participate by doing.
6. **Integration** of both meaning and form.
7. **Reflection** which gives space for learners to think over their learning process.
8. **From copying to creation** which provides opportunities for learners to use their creativity and imagination to solve real world tasks.

Ellis (2009) categorizes three approaches to TBLT according to five characteristics namely (a) the provision of opportunities for natural language use, (b) learner-centeredness, (c) focus on form, (d) the kind of task, and (e) the rejection of traditional approaches to language teaching which is illustrated in Table 4.2.

While Long and Skehan regard traditional structural teaching as theoretically unsupportable, Ellis views it as complementary to TBLT. Table 4.3 which differentiates traditional form-focused pedagogy and task-based pedagogy thereby helping to develop a clear understanding of what TBLT is.

Table 4.2 Three Approaches to TBLT

Characteristics	Long (1985)	Skehan (1998)	Ellis (2003)
Natural language use	Yes	Yes	Yes
Learner centeredness	Yes	Yes	Not necessarily
Focus on form	Yes – through corrective feedback	Yes – mainly through pre-task	Yes – in all phases of a TBLT lesson
Tasks	Yes – unfocused and focused	Yes – unfocused	Yes – unfocused and focused
Rejection of traditional approaches	Yes	Yes	No

Table 4.3 Difference between Traditional Methodology and Task-Based Methodology

Traditional form-focused pedagogy	Task-based pedagogy
Right discourse structure consisting of IRF (initiate – respond – feedback) exchanges	Loose discourse structure consisting of adjacency pairs.
Teacher controls topic development	Students able to control topic development.
Turn-taking is regulated by the teacher	Turn-taking is regulated by the same rules that govern everyday conversation (i.e. speakers can self-select).
Display questions (i.e. questions that the questioner already known the answer)	Use of referential questions (i.e. questions that the questioner does not know the answer to)
Students are placed in a responding role and consequently perform a limited range of language functions.	Students function in both initiating and responding roles and thus perform a wide range of language functions (e.g. asking and giving information, agreeing, and disagreeing, instructing).
Little need or opportunity to negotiate meaning.	Opportunities to negotiate meaning when communication problem arise.
Form-focused feedback (i.e. the teacher responds implicitly or explicitly to the correctness of student's utterances)	Content-focused feedback (i.e. the teacher responds to the message content of the students' utterances).
Echoing (i.e. the teacher repeats what a student has said for the benefit of the whole class)	Repletion (i.e. a student elects to repeat something another student or the teacher has said as private speech or to establish inter subjectivity).

Task-Based Model of Lesson Plan

Since the 1990s, a comprehensive structure for the communicative class room has been framed taking into account the importance of methodology of TBLT in which teachers implement tasks, set learners up to perform the tasks successfully and manage learners' attention to form-meaning connections (Samuda, 2001). Nunan (1989) claimed that the 'what' of teaching (course design) and 'how' of teaching (methodology) are merged in TBLT. Similarly, Kumaravadivelu (1993) argued that 'methodology becomes the central tenet of task-based pedagogy' (p. 73) since the goal is to allow learners to navigate their own paths and routes to learning. Various lesson designs have been proposed (e.g. Estaire & Zanon, 1994; Prabhu, 1987; Skehan, 1996; Lee, 2000) which have in common three principal phases: 'pre-task', 'during task', and 'post-task'. Three significant models for task-based methodology have been manipulated by Ellis (2003), Nunan (2004), Willis (1996) that differ slightly in scope but are similar in the way they encourage teachers to reflect on different methodological options around and during tasks in order to maximize learners' task performance and learning.

The Ellis Model

The Ellis (2003) model is the simplest and most adaptable for teachers and involves three optional phases: the pre-task phase, the during-task phase, and the post-task phase. Table 4.4 which showcases a frame work for designing task-based lesson elucidated by Ellis (2003) provides a clear structure for a lesson and it also allows for creativity and variety in the choice of options in each phase.

The purpose of the pre-task phase is to prepare students to perform the task in ways that will promote acquisition. Lee (2000) suggests that pre-task should highlight what the students will be required to do and the nature of the outcome they will arrive at. Dornyei emphasizes the implicit instruction and motivation. Ellis recommends four alternatives available to the teacher during the pre-task stage:

1. Supporting learners in performing a task similar to the one they will perform during the second stage;
2. Asking students to observe a model of how to perform the task;
3. Engaging learners in non-task activities designed to prepare them to perform the task; or
4. Strategic planning of the main task performance.

To utilize the during task phase effectively, the teacher has to plan beforehand the task performance options. A number of 'process options' are also available "involving both the teacher and the students in on-line decision making about how to perform the task as it is being completed" (Ellis, 2003, p. 85). Ellis details three task performance options:

1. Whether to require the students to perform the task under time pressure
2. Whether to allow the students access to the input data while they perform the task
3. Whether to introduce some surprise elements into the task

Table 4.4 Task-Based Lesson Plan: Ellis Model (2003)

Phase	Examples of options
Pre-task	• Framing the activity • Planning time • Doing a similar task
During task	• Time pressure • Number of participants
Post-task	• Learner report • Consciousness raising • Repeat task

Process options must be taken while the task is being performed. Practical knowledge and use of theory help the teachers on the spot decision about how to conduct the discourse of a task. Ellis points out the three major pedagogic goals in post-task phase:

1. To provide an opportunity for a repeat performance of the task
2. To encourage reflection on how the task was performed and
3. To encourage attention to form, in particular to those forms that proved problematic to the learners when they performed the task.

The Willis' Model

Table 4.5 illustrates the lesson plan model formulated by Willis (1996, p. 38). Willis outlines a task-based lesson plan model with three stages of pre-task, task cycle, and language focus. During the pre-task stage, the teacher identifies and introduces the topic and learners feel motivated to perform the task. The second stage, task cycle gives learners the opportunity to perform real-world tasks with teacher's monitoring and facilitation. The final stage, language focus, places emphasis on language features used during the previous stages. The teacher's monitoring during the task cycle is a kind of informal assessment since he/she provides indirect feedback (Willis, 1996, p. 169). In the Willis model, explicit grammar teaching can only take place in the analysis component of the language focus (i.e. after the task) whereas in Ellis model (2003) grammar teaching as a means of focus on form can take place in any of the phases.

Table 4.5 Task-Based Lesson Plan Model (Willis, 1996)

Pre-task

Introduction to topic and tasks
Teacher explores the topic with the class, highlights useful words and phrases.
Learning may be exposed to examples.

Task cycle		
Task	**Planning**	**Report**
Students do the task in pairs or small groups. Teacher monitors; mistakes do not matter.	Students prepare to report. Accuracy is important, so the teacher stands by and gives advice.	Students exchange or present report. Teacher listens and then comments.
	Language focus	
Analysis	**Practice**	
Students examine then discuss.	Teacher conducts practice of new words.	

In addition, Nunan (2004) has manipulated task-based framework which involves six steps for teachers to follow in a deliberate attempt to simulate receptive to productive processing. These are (a) schema building, (b) controlled practice, (c) authentic listening, (d) focus on linguistic elements, (e) providing freer practice, and (f) doing a fully communicative task. Of these three models, Willis' (1996) task-based framework is the strongest version of TBLT.

Role of Teachers and Students

The overall purpose of task-based methodology is to create opportunities for language learning and skill development through collaborative knowledge building. Ellis recommends the following principles that can be used to guide the selection of options for designing lessons:

1. Ensuring an appropriate level of task difficulty
2. Establishing clear goals for each task-based lesson
3. Developing an appropriate orientation to performing task
4. Ensuring that students adopt an active role in task-based lessons
5. Encouraging students to take risks
6. Ensuring that students are primarily focused on meaning when they perform a task
7. Providing opportunities for focusing on form also
8. Requiring students to evaluate their performance and progress

Tasks as such do not determine learning. Though learner-centeredness is emphasized by the theoreticians in TBLT, it is actually a collaborative work of both teachers and learners. Teacher's concern from the beginning of a task to the very end is very important starting from organizing the tasks, providing clear instructions, putting learners in groups, organizing turn-taking during whole class discussions, maintaining control over what happens in the class room and checking the sequence of activities for logic, coherence, and level of challenge.

During the implementation of all the three stages of tasks, pre-task, during task, and post-task, teachers address students' needs and interests by becoming facilitators. In the stage of pre-task, they play the role of selector and sequencer of tasks in keeping with learner needs, interests, expectations, and language skill levels. It is teachers who design the tasks as work plan. During task phase where work plan turns into action and interaction teacher should assume the role of interactional partner and supporter. In the post-task stage, the teacher's role is consciousness-raising by focusing on forms, examining the given text, and exposing and guiding the learners to similar tasks. Assuming the roles of 'instructor' and 'interlocutor', teachers can help students through their 'instructional conversations' to perform new linguistic features. According to Willis and Willis (2007), teachers engaging in TBLT become leaders and organizers of discussions, managers of group or pair work, motivators to engage students in performing a task and language experts to provide language feedback when needed. Thus, a teacher plays a crucial role in TBLT though it emphasizes the central role of the learner. It is true according to the famous quote "the quality of an education system cannot exceed the quality of teachers."

Students in TBLT classroom should take the leading role in their own learning as the task-based lessons are student-centered. Learners are autonomous to negotiate course content or to choose linguistic forms when performing a task. They are group participants and most importantly risk takers and innovators as they have to constantly face challenges involved in the use of target language. Therefore, both teachers and learners in TBLT classroom are responsible for the development of classroom interaction.

Significance of TBLT

Major benefits of practicing TBLT pointed out by elite researchers like Ellis, Willis, and Van den Branden, 2006, are as follows:

- Task-based work as group and pair work gives students extensive opportunities to learn, practice and develop their communicative skills in a natural way.
- It stresses meaning over form; however, it can also emphasize learning form.
- It offers learners a fertile input of target language.
- It is intrinsically motivating and helps students to build more self-confidence to use the target language both inside and outside the classroom.
- It contributes to the improvement of communicative fluency while not disregarding accuracy.
- It reaches beyond language learning offering learners the opportunity to develop self-regulation skills, problem-solving skills, intercultural competence, and social skills.

Although task-based approach presents many benefits to aid foreign language learning, it is not without some obstacles and challenges. The essential steps of designing, developing, and implementing tasks consume more time and cognitive effort to reach the goals. Hatip (2005) traces out some challenges of task-based approach saying that the drawbacks of task-based learning rely not so much on the potential powerfulness of instructional content but on problems of conducting the instruction (as cited in Tan, 2016). Task-based learning involves a high level of creativity and dynamism on the part of the teacher. Some learners employ the mother tongue when they face with a difficulty or if the group feels intolerant. There is a danger for learners to attain fluency at the expense of accuracy.

Teachers have also raised a wide range of critical concerns with a task-based approach language teaching: A task-based approach with its primary focus on meaning-making may clash with (a) official, standardized, and form-focused tests that students are supposed to prepare for; (b) the crucial importance that teachers assign to the development of explicit (grammatical) knowledge and a primary concern for accuracy of output; and (c) teachers' beliefs that the development of communicative language skills is based on the preceding development of explicit knowledge.

- Peer interaction during task-based work may lead to increased levels of noise in the classroom and to an increased use of the students' mother tongue.
- It is very difficult to implement a task-based approach in large classes.
- The emphasis on learner initiative, autonomy, and independence conflicts with more hierarchical views of the student-teacher relationship and with the role of the teacher as expert and superior.

Taxonomy of Task Types

In the view of Ellis (2009), for the effective implementation of task-based pedagogy, first of all, the teacher should have an overt comprehension of what a 'task' is. Then, he/she begins by choosing a topic, narrows it down and designs the different kinds of tasks. While developing the tasks, there will be different language needs as every individual learner possesses different personality traits, styles and strategies, aptitudes for acquisition and motivation. Therefore, teachers can set up different types of tasks which are classified according to cognitive processes as shown in Table 4.6.

Willis and Willis (2007) say that "a good task not only generates interest and creates an acceptable degree of challenge, but also generates

Table 4.6 Taxonomy of Task Types in Willis and Willis (2007)

Task types	Examples of specific tasks
Listing	Brainstorming fact-finding games based on listing: quizzes, memory, and guessing
Ordering and sorting	Sequencing ranking ordering classifying
Comparing and contrasting	Games finding similarities and differences graphic organizers
Problem-solving tasks	Logic problem prediction
Projects and creative tasks	Newspaper posters survey fantasy
Sharing personal experience	Story telling anecdotes reminiscences
Matching	Words and phrases to pictures

Table 4.7 Main Differences between Focusing on Language and Form (Willis & Willis, 2007)

Focus on language	Focus on form
• Students' initiative and needs. • Takes into account the context of the communicative activity. • Students explore what they need. • Student-centered.	• Teacher's initiative and need. • Outside the context of the communicative activity. • Teacher provides what students need. • Teacher-centered.

opportunities for learners to experience and activate as much language as possible" (p. 70). Focus on language and form depends on how tasks are graded. At this point, it is necessary to distinguish among focus on meaning, focus on language, and focus on form. Table 4.7 points out the main differences between focusing on language and form (Willis & Willis, 2007, p. 114).

Focus on language occurs when learners "pause their process for meaning and switch to thinking about the language itself" (Willis & Willis, 2007, p. 113).

Need for TBLT

For professional students, it is utmost important to develop their academic LSRW proficiency beyond grammar and vocabulary learning to hone their communicative skills. Development of receptive abilities reading and listening functions primarily in academic activities whereas that of listening and speaking becomes more essential in interpersonal production and interpretation. Listening is a critically important activity both for learners who want or need to participate in oral interpersonal (reciprocal) communication and for learners who want or need to receive information from such non-reciprocal oral sources as lectures and media broadcasts. Oral communication which is also a crucial part of second language acquisition can take many forms, ranging from informal conversation which occurs spontaneously to participation in formal meetings that occurs in a structural environment. It is not merely repeating or memorizing dialogues, but the ability to use the utterance to communicate in the real situations. It is an interactive process of constructing meaning that involves producing, receiving, and processing information (Burns & Joyce, 1997). Ur (2006) claimed that of the four language skills (listening, speaking, reading, and writing), speaking seems to be the most important, but challenging. Students find it very difficult as they have to face the audience and need preparation to produce the language. This difficulty is basically due to their inadequate frequency of speaking opportunities in the classroom. Many students undergo psychological constraints before they are prepared to speak in the foreign language. They feel so anxious that they prefer not to speak at all and deny opportunities for practice. Hence, teaching speaking involves more efforts from the teachers' side encouraging learners to talk as much as possible to convey the messages of communication.

TBLT provides more space to enhance students' confidence and it allows each student get the opportunity to experience the learning process by putting him/her in a situation in which she/he can decide alone, mainly the way of expressing and accepting experience of using the target language. As functionalism accounts for speaking phenomena, skills in speech acts help them to practice interpersonal communication. Practicing utterances in the context of requesting, apologizing, promising, denying, expressing emotions, complimenting, complaining, etc., help them to handle the communication part effectively in real-life situations. Thus, active engagement of learners in interaction, participation in communicative events, collaborative expression, modified input, feedback, and negotiation of meaning facilitates language acquisition.

Speaking Proficiency

Accuracy and fluency of language is also expected as a part of advanced academic language as well as interpersonal competence. Speaking fluency is defined as the learners' ability to produce a speech that is rapid and comprehensible. The real achievement of the learner in developing fluency is possible when he/she practices real-life time activities automatically and engages effectively in all language activities (LSRW) whether receptive or productive.

Researchers on speaking skills have recommended speaking tasks to be given to students as they enable rehearsal, feedback, and active engagement. Dealing with the importance of speaking in EFL, Stovall states that language learners need to recognize that speaking involves three areas of knowledge:

1. Mechanics (pronunciation, grammar, and vocabulary): Using the right words in the right order with the correct pronunciation.
2. Functions (transaction and interaction): Knowing when clarity of message is essential (transaction/information exchange) and when precise understanding is not required (interaction/relationship building).
3. Social and cultural rules and norms (turn-taking, rate of speech, length of pauses between speakers, relative roles of participants): Understanding how to take into account who is speaking to whom, in what circumstances, about what, and for what reason. (as cited in Malihah, 2010, p. 88)

Overall, the major contents of language pedagogy like attention to meaning, engagement with grammar, inclusion of pragmatic properties, use of authentic communication, importance of social interaction, integration of language skills, and the connection to psycholinguistic processes are involved in speaking task.

Methodology

The present study has been an attempt to gain some insights into how learners could improve their speaking fluency in the TBLT approach. It is a descriptive account of classroom experience. It has employed two major instruments for collecting both quantitative and qualitative data, namely the questionnaire and observation. The student questionnaire consists of their feedback about the advantages of task-based speaking activities. All the items of the questionnaires have been specially designed to say whether they agree or disagree marking either yes or no options. The main purpose of observation in this study is to verify the ways in applying TBLT principles in teaching. The data sets have been qualitatively analyzed.

A framework for TBL consisting three phases pre-task, task cycle, post-task (language focus) combining the features pointed out by Willis (1996) along with assessment has been deployed in the study. Learners' role is more important because their attitude to learning, level of motivation, anxiety, and learning styles have more effect than materials or methods pragmatized by teachers. No one can be forced to learn without their involvement. Therefore, the tasks have been designed in such a facilitative and enjoyable way that would stimulate a need and desire to learn. Focusing both on syllabus and day-to-day activities, students have been assigned the tasks to compete so that they can practice the language in a personalized and meaningful way. Students' college or social life–related activities which they are familiar with are designed to reduce their anxiety level. Ensuing the principles of Ellis (2003), they are given chances to tryout language and to make the best use of what they know in various real-life situations. The following tasks have been experimented with the second-year engineering students in the below order increasing the level of complexity:

Task 1: Self introduction, getting, & providing personal information
Task 2: Giving and getting directions
Task 3: Picture description
Task 4: Listening to a TED Talk & summarizing
Task 5: Short presentation

Implementation

Task I

At first, information-gap activity which involves a transfer of given information from one person to another was given. As a pair work, students were supposed to perform the task of introducing themselves to the partner. After that, they were instructed to get the personal information about the partner through all possible questions and provide information about themselves through proper response. Enough time was given to them to prepare questions and answering them in a proper way. During the task, they were motivated by the teacher to come out with more questions and it was found that task interactions probably helped learners in gaining some automaticity in speech production. Students then attempted to convey verbally the personal profile of the partner based on the details they had gathered from the pair to the whole class. At last, the teacher focused on the language part of framing wh-questions and yes/no-type questions and polite way of answering. For further practice, they were shown the video on the formal self-introduction in an interview and asked to produce a video introducing themselves formally and submit as assignment. Their performance was assessed using rubrics and feedback was provided. It is observed that students' enthusiasm heightened as they recorded their own video after so many rehearsals and their own corrective feedback.

Task 2

After their initial activity, students were assigned situations on the spot and they had to perform at the very same time. Two pairs of students were called upon to do the task of getting and giving directions from the central bus stand to reach the boys and girls hostel of Anna University, Trichy. While they hesitated, they were motivated by the teacher, and after a while they started exhibiting interest. During the pre-task, their interest was in achieving the goal looking for meaning. Directions-related words, phrases, and sentences were analyzed. Afterward, all the students were given a picture of places and buildings and asked to give direction to the places that others want to arrive at as a pair work. Actually, students were elated as they felt that they were able to develop fluency as well as accuracy.

Task 3

Next, students attempted reasoning-gap activity of picture description which involved deriving some new information through processes of inference, deduction, practical reasoning, or a perception of relationships or patterns. They were given two pictures from which they chose one to describe at once. Their speeches were recorded before and after utilizing the TBLT approach. They could not use proper lexical items they were expected to know in the beginning. They mostly focused on meaning without bothering grammar forms especially of tenses and prepositions. The teacher guided them by showing the example of how to describe meticulously focusing on actions, movements, colors, facial expressions with appropriate vocabulary and grammar forms. Words used by learners before and after were counted. This word count helped in assessing how the learners improved their speaking fluency in terms of speed production and grammatical accuracy and comprehensibility. The learners not only improved their grammatical accuracy but also the comprehensibility. They increased the number of words they uttered. In some instances, before teaching, it was hard to understand what the learner was saying. By contrast, after teaching all the clauses flowed clearly despite some grammatical inaccuracies. Learners were able to elaborate on their statements. Before teaching, the fillers were used as long pauses whereas, after teaching, the fact that they may be used as one of the strategies to maintain the flow of discourse, they were used effectively as short pause. In the expressions such as 'it means', 'I think', 'I guess', during task interactions, learners gained abilities to express their thoughts, feelings, and reactions which could be relevant for daily, oral communications.

Task 4

Students were asked to listen to the TED Talk of their own interest and choice. Though the task seemed to be a listening comprehension activity, implicitly, students were made to observe the model of doing presentation effectively. Students were required to pay attention to how the speakers keep their presentation interesting and motivating. Skehan (1996) and Willis (1996) suggested that simply observing others performing a task can help the cognitive load on the learners. This was well observed when they were asked to present the summary of the talk. Moreover, they were given enough time to plan how to complete the task in their best way. Without hesitation they were able to present, as they had the main ideas and key words and phrases.

Task 5

Students were assigned the task of giving individual mini presentation on the topic of their own choice. They were provided a wider range of options like sharing personal experiences, comparing, and contrasting, narrating, describing, or articulating their own opinions and suggesting solutions. As they were given a short period of time to prepare and also do it individually, they struggled a little bit to organize themselves. They searched for useful vocabulary and language structure. After the task, the teacher asked the students to consider the language they used. Peer reviews and self-assessments were embedded in the task designs.

Assessment

In all these activities, the teacher observed the performance of the students and provided feedback commenting on the individual performance and pointing out the common errors. Throughout the tasks, learners were supported in expressing meanings and also helped in improving linguistic forms by using 'recasts' by reformulating their speech in the correct form whenever needed. Other explicit and implicit interactional devices such as request for clarifications, questions, and advice were also encouraged. When the errors were pointed out, the students tried to avoid the errors in the following tasks. The tasks were well received by the majority of the students. They found the experience to be rewarding, intrinsically interesting, and educationally beneficial. They got involved in the task because the tasks were encouraging them to perform on real-life situations. They themselves realized their responsibility toward their learning and started to focus on language while practicing. The final performance was much improved in all the activities and learners themselves perceived that they had made significant progress with their speaking fluency.

Students were assessed by their performance on these tasks. Formal assessment using rubrics to evaluate task performance using the target language was done. Secondly, informal assessment using a self-evaluation format was carried out. Self-assessment form would enable students reflect upon their final understanding of the task and provide their feedback and attitudes toward the given task. Ultimately, this evaluation form would provide insights about four different aspects: task goal, task performance, kind of interaction, language focus, and future actions.

Results

Questionnaire has been issued to the students in the end of the course to get their feedback about task-based activities. 97% of the students have found the tasks to be interesting. 98.9% of them have admitted that these tasks on speaking activities have demanded their active participation. 99.4% have responded that tasks have enabled their confidence level. 96% have opined that these tasks have enabled their imagination and creativity. Tasks have helped them to make decisions about topics of their own to 97.7% of the students. 92% learners have felt that the tasks provide a relaxing English learning atmosphere for them. The tasks provide real-life situations for 94.9% to practice speaking. 97.7% learners have remarked that the tasks have increased student-teacher interaction. 83.4% of students have liked to be assigned similar tasks. 64.8% have preferred to work individually whereas, 86.9% have liked to work in groups. 69.5% have admitted that they had difficulties in understanding what the peers have been saying. 51.7% have had difficulty in expressing their thoughts when asked to do the task on the spot. When learners have been instructed to complete the task through preparation it has been found that it reduces their troubled feeling in mind or anxiety. 80.7% have acknowledged that they have been given right amount of time to do their task.

Conclusion

TBLT facilitates collaboration interactions and communication cultivating positive affects toward language learning. Most of the learners consistently express that they are more confident, feel secure, and casual as they communicate with their peers and also feel a sense of accomplishment.

They enjoy a certain degree of freedom when asked to select their own partners and topics. They try to produce more coherent speech. It is believed that the consistent practice of such kind of tasks would definitely lead to gradual learning of using appropriate expressions in variety of contexts. Learners themselves acknowledge improvement in their language proficiency because of the opportunities provided to practice L2 through these tasks. Undeniably, TBLT has more advantages as its principles try to balance between language use, language meaning, and language forms and to integrate all skills and elements. On the whole, it is observed that task-based activities are certainly beneficial to the ESL students and could be practiced often in the class rooms as they enhance interaction, confidence, motivation, language skills, and knowledge of the learners.

References

Beglar, D., & Hunt, A. (2002). Implementing task—based language teaching. In J. Richards & W. A. Renandya (Eds.), *Methodology in language teaching: An anthology of current practice* (pp. 96–106). Cambridge: Cambridge University Press.

Branden, K. V. D. (2006). *Task based language education: From theory to practice.* Cambridge: Cambridge University Press.

Breen, M. (1987). Learner contribution to task design. In C. Candlin & D. Murphy (Eds.), *Language learning tasks* (pp. 23–46). Englewood Cliffs, N.J.: Prentice Hall.

Burns, A., & Joyce, H. (1997). *Focus on speaking.* Sydney: National Center for English Language Teaching and research.

Carless, D. (2002). Implementing task—based learning with young learners. *ELT Journal, 56*(4), 389–396. doi: 10.1093/elt/56.4.389

Ellis, R. (2003). *Task based language learning and teaching.* New York: Oxford University Press.

Ellis, R. (2009). Task-based language teaching: Sorting out the misunderstandings. *International Journal of Applied Linguistics, 19*(3), 221–246. doi: 10.1111/j.1473-4192.2009.00231.x

Ellis, R., & Shintani, N. (2014). *Exploring language pedagogy through second language acquisition research.* New York: Routledge.

Estaire, S., & Zanon, J. (1994). *Planning classwork: A task based approach.* Oxford: Heinemann.

Hatip, F. (2005). Task-based language learning. Available online at http://www.yde.yildiz.edu.tr/uddo/belgeler/inca-FundaHatip-TBL.htm, accessed May 25, 2010.

Kumaravadivelu, B. (1993). The name of the task and the task of naming: Methodological aspects of task-based pedagogy. In G. Crookes, & S. Gass (Eds.), *Tasks in a pedagogical context: Integrating theory and practice* (pp. 69–96). Clevedom: Multilingual Matters.

Lee, J. F. (2000). *Tasks and communicating in language classrooms.* Boston: McGraw-Hill.

Littlejohn, A. (1998). The analysis of language teaching materials: Inside the Trojan Hhrse. In B. Tomlinson (Ed.), *Materials development in language teaching* (pp. 179–211). Cambridge: Cambridge University Press.

Littlewood, W. (2004). The task-based approach: Some questions and suggestions. *ELT Journal*, 58(4), 319–326.

Long, M. (1985). A role for instruction in second language acquisition: Task-based language teaching. In K. Hylstenstam, & M. Pienemann (Eds.), *Modelling and assessing second language acquisition*. Clevedon, UK: Multilingual Matters.

Malihah, N. (2010). The effectiveness of speaking instruction through task-based language teaching. *Register, 3*(1), 85–101.

Murphy, J. (2003). Task—based learning the interaction between tasks and learners. *ELT Journal, 57*(4), 352–360. doi: 10.1093/elt/57.4.352

Nunan, D. (1989). *Designing tasks for the communicative classroom*. Cambridge: Cambridge University Press.

Nunan, D. (2004). *Task—based language teaching*. Cambridge: Cambridge University Press.

Nunan, D. (2005). Important tasks of English education; Asia—Wide and beyond (electronic version). *Asian EFL Journal, 7*(3), 5–8.

Prabhu, N. (1987). *Second language pedagogy*. Oxford: Oxford University Press.

Richards, J., & Rodgers, T. (2001). *Approaches and methods in language teaching* (2nd ed). Cambridge: Cambridge University Press.

Samuda, V. (2001). Guiding relationships between form and meaning during task performance: The role of the teacher. In M. Bygate, P. Skehan, & M. Swain (Eds.), *Researching pedagogic tasks: Second language learning, teaching, and testing* (pp. 119–140). Harlow, UK: Pearson Education.

Saville-Troike. (2010). *Introducing second language acquisition*. Cambridge: Cambridge University Press.

Skehan, P. (1996). A framework for the implementation of task-based learning. *Applied Linguistics, 17*(1), 38–62. doi: 10.1093/applin/17.1.38

Skehan, P. (1998). Task—based instruction. *Annual Review of Applied Linguistics, 18*, 268–286. doi: 10.1017/S0267190500003585

Swan, M. (2005). Legislation by hypothesis: The case of task-based instruction. *Applied Linguistics, 26*(3), 376–401. doi: 10.1093/applin/ami013

Tan, Z. (2016). Benefits and implementation challenges of task-based language teaching in the Chinese EFL context. *International Journal for Innovation Education and Research, 4*(3), 1–8.

Ur, P. (2006). *A course in language teaching: Practice and theory* (13th ed.) Cambridge: Cambridge University Press.

Willis, D., & Willis, J. (2007). *Doing task-based teaching*. Oxford: Oxford University Press.

Willis, J. (1996). *A framework for task-based learning*. Essex, UK: Longman.

Part III

Listening, Speaking, Reading and Writing Skills

Chapter 5

Teaching LSRW Skills through the Test, Teach, Test (TTT) Method

P. Hiltrud Dave Eve

Teaching English in the contemporary Indian society is a challenging task especially in rural areas where students are from Tamil medium. It is not only a difficult task for the teachers but also for the learners. The ultimate goal of learning a language is to communicate efficiently, and it is well-known that no learning is possible if one doesn't get a chance to communicate well in the learned language. There is a high demand for communication in the ever-changing world. Language plays an important role in communication, and English undoubtedly is the leading and most important tool of communication around the world. Language learning is an ability, in which perfection can be attained only through continuous practice and exposure to the target language. In India, even though students have been provided exposure to English as a second language right from their primary level, there seems to be a great variance between the city-based children and the rural children getting accustomed to English language. Since 70% of the students are first-generation learners who hail from rural background, they lack guidance of English language from their parents and others.

Status of Rural and Urban Students

Lot of modules is exposed to city-based students and they have the privilege of learning English through different methods. Varieties of modules and special training is available in the schools situated in city to improve the students' listening, speaking, reading, and writing skills which forms basic skills for learning any language, whereas the village students have only limited exposure to learn English. The teaching methods in schools are theoretical and bookish, taught only in the examination point of view. Hence, the students hesitate to communicate in English even though they learn English for about 12 years. The same situation applies to city-based children also; many of them are good in English, but when it comes to communication they hesitate to speak in English. It is in the hands of the teacher to create awareness among the students in the school about the importance of English language. The teachers in rural areas should be committed and honest to equip the skills of the students. The teacher has to create encouraging

DOI: 10.4324/9781003268529-8

atmosphere for learning and practicing the language. Today, conventional teaching methods are substituted with modern techniques, which depend hugely upon media resources and teaching English with modern techniques fosters a positive attitude among the rural students to learn the language which would enable them to meet the demands of the day in a creative way.

Problems Faced by the Rural Students

There are so many obstacles faced by the rural students in the society. The primary factor is the socio-cultural and monetary background of the family. As the parents are not aware of education, and are being devoid of education, they are not able to guide their children. Parental supervision and guidance are the most importance factors in education. The rural background students lack their parental supervision that they no longer understand the significance of communication skills which plays a vital role. In most schools, English is taught like any other subject and the students are not aware of the importance of English in the contemporary competitive world. The rural students are not aware of the communication tool and the awareness is not given to them properly by the teachers. The existing methods followed in schools are not enough to train the students to communicate effectively. It is a very challenging task for the teachers to get the students involved in learning English as a Foreign Language (EFL). The lack of facilities and the level of motivation are low in rural schools when compared to city-based schools. City-based students have chances of learning language in multiple mediums. They have a lot of facilities to get exposed to English language and culture. In this line, motivation plays a crucial role in students' language learning process. Motivation is divided into two types: instrumental motivation and integrative motivation. The former focuses on the rewards and the latter focuses on the urge to be a part of the community, so strong that drives the students to pick up the language with perfection.

Effective Teaching Methods in English

There are so many methods followed by the expert for a long time, to equip the students to be familiar with reading, writing, listening, and speaking. The methods are discussed in the following sections.

Grammar-Translation Method

Grammar-translation method evolved from classical method. Grammar-translation method analyses language in different elements of language and teaches how to combine the elements. Learning is through translation from target language to mother tongue. Grammar-translation method instructs only in grammar and gives vocabulary with direct translations. This method of teaching is found to be useful and is successful even in a classroom with large number of students. No oral work takes place in the classroom, only teacher explains everything. The main focus is only on mother tongue and the target language is ignored besides providing main emphasis only on grammar rules. It does not help the students to correct the pronunciation and other necessary language skills.

Direct Method

In the 19th Century, direct method was introduced to face the shortcomings of the grammar-translation method. This method was developed as an antithesis to the previous method. It is also known as 'the natural method'. In direct method, language is learned inductively. Grammar is not taught plainly in direct method. This type of teaching focuses mainly on the target language. The learner in this method is not allowed to use his/her mother tongue. Grammar is taught inductively with focus on speaking and listening, and only the everyday language is taught. Even though it has advantages of learning over 500 of the commonest English words to be used in sentences, psychologically it proceeds from concrete to abstract, and there are many abstract words which poses difficulty for the students to interpret directly in English. More energy and time are consumed in this method. In higher classes, this method of teaching is identified to be unsuitable.

Audio-Lingual Method

The theory is based on the moto that learning a language means acquiring habits. In this method, much emphasis is given on practicing situational dialogues. The students are asked to listen and then practice the language before submitting it in written format. It is believed that the learners first speak what they listen, then read the same text, and write what they have read. Audio-lingual method emphasizes on listening and speaking skills. In this method, students play a passive role in the classroom. Very little attention is paid to communication and content. Due to these disadvantages, teachers will find ways to ensure pitfalls of the audio-lingual methods (Lake, 2013).

The Structural Approach

This approach is based on the assumption that language can be learned through grammatical structures. In Menon & Patel's opinion, "the structural approach is based on the belief that in the learning of a foreign language, mastery of structures is more important than the acquisition of vocabulary". Teaching is done in situations and speech is given importance, and reading and writing are also included. The advantage of this method is that it emphasizes on four skills such as listening, speaking, reading, and writing. However, the demerit of this approach is that it is suitable for primary students and not for secondary students. Continuous teaching of language structures and the repetitions make the atmosphere dull and monotonous. The students find that reading and writing ability is neglected.

Suggestopedia

This language teaching approach focuses on the learners to help them to communicate more effectively and correctly in real-life situations. According to Xue, it focuses on how to deal with the relationship between mental potential and learning ability to use in teaching speaking for young language learners (as cited in Apriana & Islamiyah 2011). It involves important items like suggesting, thanking, inviting, complaining, asking for directions, etc.; there are many advantages and disadvantages of this method. In this method, classes are held in ordinary room with comfortable chairs, a practice that may also help them relaxed. The disadvantage of this method is that only limited media will be used. There are some activities in suggestopedia that the teacher needs to be expressive.

Total Physical Response

Total physical response, also known as TPR, is an approach which follows the idea of learning by doing. Repetitive actions such as 'Stand up', 'Open your book', will be given to beginners to learn language. The most important skill is aural comprehension and everything else will follow naturally later in the most TPR. This method of teaching is fun and easy, and it does not require a great deal of preparation on the part of the teacher; it is a good tool for learning vocabulary. The disadvantage of this method is that it does not involves innovative methods. It is limited and combined with other approaches. It is not flexible to teach everything.

Communicative Language Teaching (CLT)

Communicative language teaching approach focuses on the learners to help them to communicate more effectively and correctly in real-life situations. It involves most important items like suggesting, thanking, inviting, complaining, and asking for directions. The disadvantage of this method is that it pays insufficient attention to the context in which teaching and learning takes place. This method focuses only on fluency but not accuracy. The approach does not focus on error reduction, but it creates a situation where the learners use their own mind to solve communication problems.

The Silent Way

Learner autonomy is given prominence in this silent way method of teaching. In this method, teachers act only as facilitators; their role is encouraging students to be more active in learning. The main advantage of this way of teaching is for the teacher to say very little, so the students can take control of their learning. There is a big emphasis on pronunciation that are constantly drilled and recycled for reinforcement. The teacher evaluates their students through careful observation, and it's even possible that they may never set a formal test as learners are encouraged to correct their own language errors. Learning through problem solving is attractive and it fosters creativity, and increases intelligent potency of the students. The silent way is criticized as a harsh method and it put the learners at stake.

Task-Based Language Learning

The main objective of this approach is task completion. In this method of teaching, relevant and interesting tasks are set by the teacher, and students are expected to draw on their pre-existing knowledge of English to complete the task with as few errors as possible. There are numerous critics that disapprove components of the TBL teaching method and framework.

The Lexical Approach

The lexical syllabus is another name for lexical approach. It is based on computer studies, and on identifying the most commonly used words in the language. It mainly focuses on vocabulary acquisition and teaching lexical items. It is based on the frequency and use. In this approach, great emphasis is on authentic materials and realistic scenarios for more valuable teaching and learning. These approaches provide as an insight into the traditional methods of teaching, and their merits and demerits.

Literature Review

In recent educational setting, language testing plays a prominent role. Brown (2004, p. 3) in his book explains that a learner's ability, knowledge, or performance in a particular domain are examined in test. Test is a method which measures the understanding of a person's knowledge in a particular domain.

> Testing is not only a method for test-takers to design a quality test but also measures a test-takers' performances in a setting instrument. UR (2005, p. 33) has given four reasons for testing the students as follows,

- The teacher gets the full information of the students whether he/she understands the text, and know what to do next, whether to teach ahead or revise the same thing using her teaching methodologies.
- The students also get information about what and how much they learned, whether to go ahead or revise lessons.

- The students can place them in proper position how much they understand the syllabus.
- The teacher knows how much successful he or she is.

(as cited in Paudel, 2018)

Testing is used to get information about teachers and students on various aspects of teaching learning process. It supports the stakeholders to judge whether the teaching learning process is effective or not, and on the basis of information obtained from testing, some effort can be made to improve the effectiveness of teaching learning process.

The nature of teaching and testing has been shifted according to the changes in teaching methods. Richards (2001, p. 3) summarizes the relative supremacy of the teaching methods in education during the past centuries. Testing is not taken as the part of language teaching even with the emergence of communicative approach. This fact has created a great challenge for many language teachers to find the typical way of testing students' ability or progress in terms of their functional ability in the EFL context.

Now, with the development of task-based instruction and content-based instructions, the concept of testing has been shifted and it has become an integral part of language teaching. Showing the linkage between teaching and testing, Rudman (2004) mentions that with the development of discovery approach in teaching, testing has remained as an integral part of teaching. He further mentions four points to show the linkage between teaching and testing:

- In the beginning, testing has been considered as a useful tool, he means that a language teacher gains knowledge of students' performance in new instruction. The results of test help the teacher to plan, review, and identify the potential issues to be faced.
- Testing method helps the teacher as an aid in making decision like grouping students in the class room. It also helps the teacher to specify his instructional objective and assigning specific instruction to particular group.
- Testing can be used to diagnose what individual student knows. A triangulation of several kinds of data drawn from various types of tests like aptitude test, observation of behavior, teacher made quizzes, etc., are helpful to find out the current level of the students. Teaching can help the teacher determine the pace of the classroom.

Teachers tend to use tests prepared by themselves more often than any other types of tests to monitor what has been previously learned. Tests cannot be separated from testing as testing is a tool to get teaching improved. At the same time, to identify the exact performance of the students, a test needs to be designed from the items that has been taught in the classroom. Congruently, Widyantero (2012) argues that the close link between teaching and testing can be seen from the fact that a test cannot be acknowledged as valid, if it does not measure what it is supposed to measure, 'what' must be related to the materials taught by the teachers. Similarly, Walvoord and Anderson (2010) have stated that a language teacher needs to keep in mind that language testing can have a great effect on their students. This will affect how students learn, what they focus on, and how much time they spend. Testing, not only tests the students' and teachers' performances but also influences the students' motivation to learn as well. A language teacher should create environment, which encourages students to learn. Patel and Jain (2008) argue that some parties are concerned about the result of language testing, such as the school principal, cottage chief, teachers, students, and their parties. These parties need to know how many students have been benefited from their study, and then need to take a decision.

Significance of Test, Teach, Test Method at Present Era

This approach encourages the teacher to give an assessment before teaching and then provide another assessment at the end of the lesson. Since the students nowadays loves challenges, teachers might as well use this approach, just to make sure that content is equal with the assessment that one would provide.

Unlike other methods, it claims that a teacher must test students' prior knowledge of the target language for two reasons: one is to find out their current level of knowledge and the other one is to explore the particular problem that they have so that the teacher can plan to teach the target language systematically. Test-Teach-Test (TTT) method helps the teacher make his/her input comprehensible. Krashen concedes:

> The best methods are therefore those that supply comprehensible input in low anxiety situations, containing messages that students really production in second language, but allow students to produce when they are ready recognizing that improvement comes from supplying communicative and comprehensible input, and not from forcing and correcting production.
>
> (as cited in Schutz, 1998)

To make language input effective/efficient and comprehensible, a language teacher needs to know the current level of the learners so that he/she can plan for the language input which is i+1; a level ahead of the learners.

The TTT method can be a means to decide i+1. Talking about the relevance of this method, Hadfield (2014) has mentioned two advantages; beginning with a test allows the teacher to listen to the students and know what language they know already and what is necessary to teach them. This avoids teaching the language they know already. Similarly, it makes students aware of the gaps in their knowledge and may make them more receptive to the new language input. Bondjema explains that TTT is a useful approach as it enables teachers to identify the specific needs of the learners concerning a language area and address this need suitably (as cited in Paudel, 2018). In the same context, Woodward writes,

> In terms of opportunities for learning, the first 'test' stage offers students a chance to try to remember and use what they have remembered. The 'teach' stage may offer exposure to new language and some chances to notice features of language and the second 'test' could give the chance for use and refine level.
>
> (as cited in Paudel, 2018, p. 20)

It is deciphered from the above method that TTT is a student-centered teaching technique in which learners learn language by doing themselves. Getting insight from the above-mentioned method, there are some points given by Pitmbar Paudel (2018),

1. Teacher administers, tests, and analyzes the prior knowledge of the students in a particular text area.
2. Students first use their knowledge and answer the questions.
3. After the results, teacher centers his/her target language lesson on the results of the first test.
4. Students are made aware of their knowledge of the target language by filling the missing part of the target language (TL), and are encouraged to practice it fully to understand it.
5. Learners are given another text and are asked to find out the solution based on the model answers. The teacher plans a lesson and facilitates them and involves them in more practice.
6. Students are tested again which allows teacher to monitor and observe the target language used by the students. This stage can then be followed with delayed error-connection focusing on how students have used the new language, and highlighting the differences/improvements from the first test stage.

Advantages of Test-Teach-Test Method

TTT lessons give tangible evidences to upgrade the students. When the test items have been compared, it has been evident that upgrade is seen by the students that they can understand their stage in learning the language. TTT procedures encourage student-centered approach, as the students practice the patterns of language based on the pretest, and help them to develop and learn more on the language part in the final test. TTT allows positive comments and self-error correction. Encouraging error as a learning tool makes the students to learn new things in the next lesson and it also helps them to learn new vocabulary. The TTT method also acts as a motivator for the slow learners, as the progress is seen by them in each test. It is a method of structuring the lessons and in the initial test stage, the teacher introduces the context and observes the students' performance, in the second stage, accuracy is attained, and in the third stage, fluency is achieved in the target language.

Methodology

The study was designed as an experimental research under quantitative enquiry. The data was collected only from primary sources to find out the effects of TTT method in English classes. The total population of the study was all the first-year students of As-Salam College of Engineering and Technology. The students were divided into two groups: experimental and control groups. The experimental group students were facilitated with the TTT method. The tool for data collection was test items: pre-test and post-test (see Appendix A and B). A pre-test was administered to identify the proficiency level of the students in the beginning of the study. At the end, post-test was conducted to assess the students' level of improvement, and the data were statistically analyzed using SPSS.

Participants

The study has been conducted among the first-year students of As-Salam College of Engineering and Technology situated in Thirumangalakudi, Aduthurai, Tanjore district, Tamil Nadu. These students are first-graduate students with less knowledge in basics of English. Being educated in rural schools, they have been identified to be inadequate in English language proficiency that they are not aware of simple structures of language. It is a challenging task to teach them language; however, I have employed the TTT method to teach English language to the group.

Data Analysis

This study investigated the efforts of the TTT method in English classes. The collected data was analyzed, interpreted, and discussed based on the hypothesis that the implementation of TTT method in ESL classroom enhances students' language, reading, and writing skills.

Results and Discussion

The descriptive analysis and one-way ANOVA has been computed to analyze the proficiency level of students in the entry and exit level.

Table 5.1 exhibits the results of the pre-test. The results in Table 5.1 indicate that means of the experimental group at the entry level are 7.14 in language, 10.36 in reading, and 5.40 in writing. This depicts that the

Table 5.1 Descriptive Statistics of Pre-Test

Components	Mean	Std. Dev.	Std. Error	Max.	Mini.	Sum.	Median	Range	Variance
Language	7.14	2.391	0.286	14	1	500	7.00	13	5.718
Reading	10.36	2.808	0.336	17	4	725	11.00	13	7.885
Writing	5.40	2.595	0.310	14	0	378	5.00	14	6.736
Total	7.63	3.309	0.228	17	0	1603	7.00	17	10.951

Table 5.2 Descriptive Statistics of Post-Test

Components	Mean	Std. Dev.	Std. Error	Max.	Mini.	Sum	Median	Range	Variance
Language	9.61	3.196	0.382	15	0	673	11.00	15	10.211
Reading	15.30	1.536	0.184	18	8	1071	16.00	10	2.358
Writing	8.59	3.581	0.428	15	0	601	9.00	15	12.826
Total	11.17	4.141	0.286	18	0	2345	12.00	18	17.144

students' level of language proficiency is low, creating a need for the teacher to improve their proficiency level. It also guides the teacher to analyze the appropriate and necessary needs of the students and the innovative methodology to be implemented in the classroom to enhance their proficiency level.

Table 5.2 presents the results of the post-test. The results in Table 5.2 reveal that the means of the experimental group at the exit level are 9.61 in language, 15.30 in reading, and 8.59 in writing. These results reveal that there is improvement in the performance of the students from pre-test to post-test. This implies that the facilitation of the TTT method in the ESL classroom has improved the students' language, reading, and writing proficiency.

In addition, Table 5.3 represents the results of one-way ANOVA. The significant value in Table 5.3 exhibits that there is a significant difference in the performance of the experimental group in pre- and post-test. This shows that the TTT method has an impact in enhancing the students' language (vocabulary, grammar, synonyms), reading (comprehension), and writing

Table 5.3 One-Way ANOVA between Pre- and Post-Test

	Sum of Squares	df	Mean Square	F	Sig.
Between Groups	885.324	2	442.662	65.290	0.000
Within Groups	1403.443	207	6.780		
Total	2288.767	209			

skills (paragraph and essays). Paudel (2018) has also tested the TTT method in Nepal with ESL students as samples and has found that this method has been beneficial especially for the rural students. Congruently, the findings of the present study imply that the TTT method has enhanced the students' language skills.

Conclusion

The ESL students, especially rural students, have been identified to have lesser exposure to English language in comparison with students from city schools. This affects the rural students' English language proficiency as language learning mostly depends on learners' socio-cultural knowledge of the target language. This has resulted in the need for the rural students to be facilitated with appropriate exposure and innovative teaching methodology for achieving language proficiency. In this line, the present study has employed the TTT method with the focus to improve the students' language, reading, and writing proficiency. The analysis of the data has revealed that the experimental group students, who have been inadequate in the pre-test, have performed well in the post-test. This expounds that the implementation of TTT method in the classroom has enhanced the students' language proficiency. The findings have validated and reflected the predominance and impact of the TTT method in effectively and successfully developing the language proficiency of the rural students. The present has limited itself to rural students and has focused on the language, writing, and reading skills; however, the future studies could explore the TTT method in integration with other language components and also with different samples.

References

Apriana, A., & Islamiyah, M. (2011). A teaching method: Suggestopedia. *Novaekasari09: Wordpress*. Retrieved from https://novaekasari09.wordpress.com/2011/06/12/a-teaching-method-suggestopedia/

Brown, H. D. (2004). *Language assessment: Principles and classroom practices*. London: Longman.

Hadfield, J. (2014). *The TKT Teaching Knowledge*. Retrieved from http://www.slideshare.net/jehny.yon/the-tkt-teaching

Lake, W. (2013). *Audio-Lingual Method of Teaching English (Electronic Version)*. Retrieved from http://blog.about-esl.com/audio-lingual-method-teaching-english/

Patel, M. F., & Jain, P. M. (2008). *English language and teaching methods, tools and techniques*. Jaipur: Sunrise Publishers.

Paudel, P. (2018). Use of test-teach-test method in English as a foreign language classes. *Journal of NELTA Surkhet, 5*, 15–27.

Richards, J. C. (2001). *Curriculum development in language teaching*. Cambridge: Cambridge University Press.

Rudman, H. C. (2004). Integrating testing with teaching. *Practical Assessment, Research and evaluation: A Peer Reviewed Electronic Journal, 9*, 1–4.

Schutz, R. (1998). *Stephan Krashen's theory of second language acquisition.* Retrieved from http://www.sk.com.br/sk-krash.html.

Ur, P. (2005). *A course in language teaching.* Cambridge: Cambridge University Press

Walvoord, B. E., & Anderson, V. J. (2010). *Effective grading.* San Fransisco: Jossey Bars.

Widyantero, A. (2012). Transferability: A missing link between language and language testing. Indonesia. *FLLT 2013 Conference Proceedings, 1,* 795–801. Retrieved from http://www.litu.tu.ac.th/journal/FLLTCP/Proceeding/795.pdf

Chapter 6

Integrating the Sub-Skills of LSRW for ESL Learners

R. Vasanthan and R. Nandhini

The four language skills, listening, speaking, reading, and writing (LSRW) also known as the four core or macro skills of language, are a group of competencies that equip the learners to understand and generate spoken language for correct and successful interpersonal contact and communication. Each of the four skills has a set of sub-skills and skill activities. The macro skills refer to the primary and largest skill set relative to a particular context.

Listening skill is the ability to understand the main ideas of speeches in the target language. It is the skill to demonstrate an emerging awareness of culturally implied meanings beyond the surface meanings of the text. It is a complex, active process in which the listener must discriminate between sounds, understand vocabulary and structures, interpret stress and intonation, retain what has been gathered in all the above, and interpret it within the immediate as well as the larger socio-cultural context of the utterance (Vandergrift, 1999). Listening is the first and foremost language mode that children acquire which provides the basis for the other language arts (Lundsteen, 1979).

Speaking skill is the ability to be understood without difficulty and converse in a clear and participatory fashion. It is the skill to initiate, sustain, and bring closure to a wide variety of communicative tasks, to narrate and describe concrete and abstract topics using sustained, connected discourse. In almost any setting, speaking is the most frequently used language skill. Speaking is used twice as much as reading and writing in our communication (Rivers 1981). Speaking is closely related to listening as two interrelated ways of accomplishing communication. According to Oprandy (1994) and El Menoufy (1997), every speaker is simultaneously a listener and every listener is at least potentially a speaker.

Reading skill is the ability to easily follow the essential points of written text. It is the skill to understand parts of texts which are conceptually abstract and linguistically complex. Goodman (1967) defines reading as a selective process which involves partial use of available minimal language cues selected from perceptual input on the basis of the reader's expectation. As this partial information is processed, tentative decisions are made to be confirmed, rejected, or refined as reading progresses.

DOI: 10.4324/9781003268529-9

Writing skill is the ability to address a variety of topics with significant precision and detail. It is the skill to write competently about topics relating to particular interests and write clearly about special fields of competence. It involves the ability to organize writings with a sense of theoretical structure. It is a process that involves few distinct steps in producing a finished piece of writing. Though every writer follows his or her own writing process, educators have found writing to be a complex combination of skills, and have stated that by focusing on the process of writing, almost everyone can learn to write successfully.

Listening and speaking are brain input skills while reading and writing are brain output skills. Listening and reading are known as 'receptive' skills while speaking and writing are known as 'productive' skills. To develop the core skills, a set of sub-skills are needed to be intrinsically set in place since the core skills cannot stand alone in the language learning or application process. The validity of sub-skills being pertinent to all the four major skills can be understood by the statement of John Field (2008), "A conventional listening comprehension lesson simply adds yet another text to the learners' experience; it does little or nothing to improve the effectiveness of their listening or to address their shortcomings as listeners." The effectiveness in the listening mentioned here can be improved with specific tasks in sub-skills.

Sub-skills are basically like building blocks for the main skills of learning and mastering English language. In order to master the target language, both the learner and the teacher need to contextualize on the sub-skills to develop and see progress in the set language. There are many sub-skills of the core language skills which can be the stepping stones toward understanding the English language. Sub-skills are considered helpful for diagnostic purposes (Alderson, 2005; Goh & Aryadoust, 2014). Language sub-skills are of varied intensity and they vary in their application levels depending upon the language ability/level of the learner; however, all play a major role in language acquisition. For instance, vocabulary as a general speaking and writing sub-skill, is known to be the basis of language and to be never underestimated in learning a target language, as it gives the ability to the learners to not only comprehend words, but to activate them as well (Hoshino, 2010; Faerch & Phillipson, 1984). The sub-skill of grammatical structures is viewed both as knowledge and ability that help language learners in the tasks of sentence formation and creation of spoken and written texts (Richards & Reppen, 2014). Pronunciation as a vital speaking sub-skill of English language learning helps learners to use vocabulary and grammatical rules communicatively and is shown to be mostly effective for the improvement of segmental and supra segmental features of the English language (Kennedy & Trofimovich, 2010).

Though language teaching is designed and trained on four major skills, it is necessary for the learners to have a comprehensive picture and command over essential sub-skills for superior application. The awareness of one's competence with regard to various sub-skills will help to achieve language ability in the classroom along with technological support and other sources, which is the need of the hour. Learners' language can be enhanced if they have the aptitude to understand the nuances of the language and are also taught the potentials that may not be directly related to the main skills of the language. Thus, it is evident that learners' insight toward the sub-skills is required to be developed in the application and usage of the four core language skills. As a result, this chapter presents the activity-based teaching methodology with regard to the potential sub-skills that can eventually improve the language skills of learners in the long run.

Theoretical Background

The structural division of language into four macro skills has been insti-gated mainly by teaching-learning objectives based on the opinions of Bloomfield (1933). The LSRW skills are not exclusive or independent sepa-rate units. All the skills consist of similar and different sub-skills, which are all integrated. However, the traditional English as a Second Language (ESL)/English as a Foreign Language (EFL) courses put heavy emphasis on the isolated approach of teaching LSRW separately. It has been inferred that focusing on more than one skill at a time can be instructionally impossible (Oxford, 2001). Likewise, several studies rely on language skill segregation as reflected in traditional ESL/EFL programs that offer classes focusing on segregated language skills. In such cases, courses are presented separately, i.e. LSRW skills are taught isolated from each other. In the segregated-skill approach, the mastery of discrete language skills such as reading and speaking are seen as the key to successful learning, and language learning is typically separated from content learning (Mohan, 1986). The traditional approach to teaching English teaches the LSRW separately by stressing on skill orientation and rote memorization (Su, 2007). Peregoy and Boyle (2001) demonstrate language learning strategies as one skill being related to only another. A group of linguists, known as the formalists, have focused only on reading and writing skills as the basic canons of language learn-ing (Sanchez, 2000). However, it can be confusing or misleading to believe that a given strategy is associated with only one specific language skill. Many strategies, such as paying selective attention, self-evaluating, asking questions, analyzing, synthesizing, planning, and predicting, are applicable across skill areas (Oxford, 1990). Individual language skills reinforce each other in the process of language acquisition (Brown, 2001). We speak, in part by modeling what we hear, and we learn to write by examining what we can read. Integration of skills and sub-skills would make it easier for learners in the natural process of language development. Natural language skills integration is inevitable, and in the language classroom, skills need to be practiced in integration (Richards and Rodgers, 2009). Regarding this, Harmer (1997) states that receptive skills (LR) and productive skills (SW), being two sides of the same coin, cannot virtually be separated by reason of the fact that one skill can reinforce another in a number of ways. It is through skills integration that the learners gain the ability to use authentic language and use it for real interaction. They can get to higher levels of skill proficiency through this approach (Oxford, 2001). In actual language use, the LSRW skills and their sub-skills are rarely used in isolation. The advantages of integrating the skills far outweigh segregated-skill instruction in improving teaching for English language learners. This chapter delves into an integrated approach of the sub-skills of LSRW for ESL learners. Common sub-skills strategies can help weave the core skills together. Teaching students to improve their learning strategies in one skill area can enhance performance in all language skills (Oxford, 1996). This can be done by the incorporation of integrated sub-skills in the language teaching and learning. Most works on English language teaching concentrate on the core skills of the language to develop upon its application and usage. There is a huge research gap in investigating the language learners' perceptions of their problems, needs, and concerns in learning English. Studies on language skills and vocabulary usually do not yet match the need of the learners since the focus is mostly on the core skills and no substantial effort is given to sub-skills, though awareness and mastery of the sub-skills is exactly what is needed by both students and teachers to develop the target language. The existing language studies mainly concentrate on the importance of the four main skills, and few delve into the details of how sub-skills need to be incorporated to make the learning process easy and less demanding. For instance, most current studies encourage students to practice listening to develop the skill without giving the details of any specific tasks. Instead, if learners are asked to listen for a specific gist or detail, the task becomes more focused and easier for the learners. The listening sub-skills can create favorable situation in this context with extensive and intensive listening to engage the students naturally.

Sub-Skills in the Engineering Curriculum

Normally, in English language teaching, particularly in engineering colleges, the emphasis tends to be on the main skills where students mostly listen, read, and write while teachers mostly speak. At times, the language skills are even taught as separate entities to the students. The hitch of this teaching methodology is the uncertainty toward the learners' comprehension of the language skills. This is a serious reason why many engineering students fail to warm up to language learning in particular, and thereby face learning loss in their studies as well as loss of better prospects in their individual or career development. A thorough study and experimentation of integrated sub-skills would make language learning undemanding and thereby make up for the lapses previously observed among engineering students. As a cluster of sub-skills are associated with the core language skills (LSRW), these sub-skills, which are a group of particular capabilities, can assist language learners to attain proficiency in each of the four skills. Engineering students in particular would benefit from approaching language learning through integrated sub-skills as it would be simple to follow compared to the regular way of learning language. Enhancing language skills by integrating the sub-skills will prove beneficial for engineering students in maintaining records, assessing projects, evaluating data, writing reports of experiments, etc. Hence, the students are required to develop a wide range of sub-skills in the language classroom to enhance their language skills. Language sub-skills are important in expediting the process of language learning. The sub-skills are mostly integrated and the level of mastery over them determines the level of efficiency and duration in the learning process of the target language. This study facilitates the sub-skills of English language in the ESL classroom through designed activities and applications to promote the language skills of the students. On the account of addressing the need for sub-skills in the engineering curriculum, the present study has framed the following research questions:

Research Question 1: Whether the conscious mastery over a set of sub-skills can lead to the development of the four language skills?

Research Question 2: Whether the nature of the sub skills, i.e. basic, foundational, or advanced, will determine the level of mastery in the main language skills and level of application?

Research Question 3: Whether the awareness of the learners' strengths and weaknesses in the language sub-skills can help both students and teachers to identify the requirements and teaching strategies to learn English more effectively?

Implementation

Since language pedagogy, tests, and theories are interrelated, it is necessary to identify the sub-skills to understand their complexity. The sub-skills mentioned in the literature of applied linguistics have been studied. The manuals of popular international EFL/ESL tests like Test of English as a Foreign Language-Internet-Based Test (TOE-FL IBT), International English Language Testing System (IELTS), Cambridge English Proficiency (CPE), Pearson Test of English (PTE) Academic, Canadian Academic English Language Assessment (CAEL) have been consulted and analyzed. The most frequent sub-skills of LSRW documented in applied linguistics literature, EFL/ESL proficiency tests, and course-book tasks have also been referred.

Instruments

The LSRW sub-skills have been facilitated in the classroom through specific activities and techniques. These activities and techniques have been designed with the help of ELT experts. Each skill is sub-divided into several sub-skill activities with emphasis on the core language skills. The study has framed a variety of activities representing a range of possibilities that could assist language learners to master the target language through mindful mastery over specific sub-skills of the main language skills. The participants have been given a range of activities and techniques to select and work on their choice of order. It is up to the individual student to decide which sub-skills of LSRW he/she wants to master in the given time and facilities. Participants are not restricted by the given suggestions and can use as many or as few as felt necessary. The sub-skills activities are based on the varied teaching and learning experiences of the authors who are in the teaching profession for almost two decades. These ideas are primarily aimed at students learning English as a second language.

Participants

Two groups of students, 30 in each, belonging to Mechanical Engineering and Computer Science and Engineering branches from University College of Engineering, BIT Campus, Tiruchirappalli, were taken as samples for this study. The Mechanical Engineering students were allocated to Group 1 and Computer Science and Engineering students were assigned to Group 2. Each skill was taught to the participants in common with a set of sub-skills to find out the effectiveness of the respective sub-skills in language acquisition.

Research Design

The research is designed on a set of activities for every selected sub-skill with clearly laid down applications and techniques for implementation.

Table 6.1 contains the integrated sub-skills of LSRW selected for this study, on the basis of which specific applications, activities, and techniques have been drafted and implemented in the course of the study.

Research Application, Activities, and Techniques

The following paragraphs outline the research design followed in the study, whereby the selected sub-skills for each of the four core skills are designed with specific application mode, followed by different activities and strategies to attain the sub-skill in focus.

Listening Sub-Skills

Under listening, six specific sub-skills have been selected for the study with diverse tasks.

- The tasks on intensive listening for detail or specific information have been facilitated with the intention to make the students focus on individual parts of the listening text and phonological features of speech.

Table 6.1 Select Sub-Skills of LSRW in Integrated Format

S. No.	Main skills	Sub-skills
1	*Listening* (integrated with speaking & writing)	i. Intensive listening for detail/specific information/discriminating similar sounds/dictation/main idea ii. Extensive listening iii. Deducing meaning from context iv. Task listening (listen and do something)/identifying an error in a transcription v. Listening for gist, global understanding vi. Predicting/predicting the end of the continuation of a message or history
2	*Speaking* (integrated with reading, writing & listening)	i. Story building ii. Role play iii. Discourse marking/paraphrasing iv. Intonation v. Fluency vi. Conversation/appropriateness in choice and range of words, pronunciation, grammar, and vocabulary
3	*Reading* (integrated with listening, speaking & writing)	i. Reading for detail/identifying a referent word in a text ii. Summarizing iii. Skimming iv. Note-taking/making v. Local comprehension vi. Global understanding/comprehension/reading for gist vii. Predicting/informed guesses viii. Scanning
4	*Writing* (integrated with speaking & listening)	i. Planning/structuring/forming/linking ii. Describing a picture/series of pictures iii. Fill in the blanks iv. Paraphrasing v. Organizing/appropriate layout vi. Writing to a situation vii. Proofreading viii. Summarizing

In addition, the purpose of this intensive listening is to enhance the students' comprehension on the meaning of the text, and on the way the meaning is conveyed in the text or speech. The activities implemented for this sub-skill comprise students listening to a partner or teacher speaking for a class memory quiz/detecting mistakes/specific information. To vary, they also have listened to articles or audio newspapers or stories to retell. After listening for particular details in a text, students gave quick responses to questions either in spoken or written form. In groups and individually too, they have listened to texts varying from pop songs to nursery rhymes and rock songs to folk songs. In addition, they have also listened to a weather report to discover and share about the local weather. Few words (fan/pan/risk/wrist) or sentences (Birds flying in the blue sky with trees and water below/A house by the mountainside) have been shared and students need to repeat correctly or show the correct picture. They have been given the option to write their answer or draw the picture as desired. In groups and in pairs, the pictures are shown and the class or partner is asked to say the correct word or sentence.

- The activities on the sub-skill of extensive listening are provided with the intent to facilitate the students to gain an overall understanding with focus on the informational content rather than over-concentrating on individual words and sentences while listening to a text. This is projected to make them understand spoken language in real-world contexts by listening or involving in massive amounts of text which they can understand easily, as well as to build word recognition speed and better use of grammar points and collocations. Extensive listening activities are provided at selected levels to help the students improve automatic processing of language which allows the working memory to concentrate on comprehending the listening material. The participants have been given a topic of their choice and interest or something they are familiar with. They have listened for main ideas, not details. They have tried to understand the important points of the listening text, video, or audio recording and find out the sources that offer the basic information. Participants have also been made to listen to stories and biography information about famous international sports or film stars (M.S Dhoni, Lionel Messi, David Beckham, Will Smith, Tom Hanks) or English reviews of movies or TV shows and write it or talk about it.
- The tasks on the sub-skill of deducing meaning from context are employed to draw the students' individual interpretive skills, appropriate inferences, and to help them comprehend the listening text and appropriately tackle it. The method adopted for these tasks has involved making the participants to listen in order to get the meaning of unknown words by using the information around it. They have also listened to various video tapes and microphones.
- The assignments on the sub-skill of task listening (listen and do something) or identifying an error in a transcription have been executed

with the intention to make students focus on listening to a particular information or order and apply it. The activities for this sub-skill have entailed participants listening to a conversation or story after which they answered to questions requiring one-word answers/correct count of words or proper nouns, adjectives, etc., to identify true or false or to say yes or no. They have also been asked to draw pictures in a particular order as heard. Another activity has been conducted, which is listening to a conversation or story and retelling in one's own words to the next person or class, for which students took notes and have executed the activity correctly. Games like 'Simon says', repeating flight or train announcements and Chinese whispers have also been played.

- The tasks on the sub-skills of listening for gist and global understanding have been intended to make the participants concentrate on listening to something to get a general idea of what it's about. In addition, the purpose of listening for gist and global understanding is to enhance students' attention to extensive listening for skimming to get a general idea about a topic. The methods adopted for these sub-skills have engaged the learners by listening to a summary of the day's news on record; the radio, tapes, or CDs, after which the understanding of the material was shared or written down.

- The activities on the sub-skill of predicting are specifically chosen to make the participants focus on predicting events or details, like the end of the continuation of a message or history, or anticipate the content of the text to say what is about to be heard before listening. The methods applied to execute this sub-skill have involved activities where the students have been arranged in groups before they listened to a text and have been asked to predict the answers to listening tasks where they have to pick out detailed information. For example, they have been asked to guess the missing information in sentences like "A stitch in time saves...." Students enjoyed this competitive element and it's always interesting to see who has made the best predictions. In another exercise, students have to identify who someone is speaking to on the phone (e.g. a shopkeeper/friend/parent/brother/sister/cousin/an architect/a builder). A grid on the board has been drawn and students have been asked to predict the vocabulary, situation, and tone of voice for each of the possibilities in teams.

Speaking Sub-Skills

For speaking, six sub-skills have been chosen for the students to learn.

- The tasks on the sub-skill of story building have been framed to facilitate the participants' ability to grasp the sequence events and vocabularies with words like, soon, shortly, then, later, next, afterwards, after that, etc., in retelling events, proceedings, and conditions in own words.

The assignments for this sub-skill involved giving a picture (from a string of pictures) to each participant to examine and speak about it. Then the participants have been regrouped and each individual talked about their own picture and the group arranged the order. The string pictures have been shown one by one and key words have been extracted from them and written on the board. To provide a basis for guided writing, the participants have been asked to copy the words on the board to frame more sentences about the pictures and to retell about them in their own words or using those on the board.

- The activities on the sub-skill of role play have been intended to encourage the students to use language in speaking, adapting to particular information and create awareness that language can be used to say what we want. Four activities have been carried out to assess the effectiveness of the aforesaid sub-skill in strengthening the speaking skill at large. In activity 1, a student has been asked to imagine he or she is a doctor/movie director/shopkeeper, etc. To guess the job, the other students have to ask him/her about his/her everyday activities. Activity 2 has been carried out with the class as a group of friends, and they have to come to a decision where to go for a picnic. Each student has a card with places they would not like to go. Using the information from the cards, they have to agree on a picnic spot and share with the class. Activity 3 has asked the students to debate on various topics like 'Pollution', 'Global Warming', Environmental Change', etc., wherein students have been assigned the role of villagers, landlords, factory owners, politicians, scientists, etc. In activity 4, students in pairs have performed regular conversations in places like a pet shop, plumbing agencies, post office, bakery, hospital, toy shop, etc., for given situations like enquiry about a particular pet, making appointment with doctor, putting order for birthday cake, complaints, requests for home service, etc. Students' works on plays have also been acted out.
- The assignments on the sub-skills of discourse marking and paraphrasing have been applied in this study to help the students support argument or viewpoint to maintain a consistent style, and to avoid the usage of lengthy quotations from the original text or conversation. The method applied here has involved the students to practice using words/ phrases which organize a talk (e.g. firstly, secondly, on the other hand, to summarize, etc.). Activities with discourse makers are planned and students have been required to use them appropriately. The participants have been made to read and make notes and find different terms to put texts into their own words, rewording messages from various sources in an assured and clear way.
- The next sub-skill under speaking skill taken up for this study is intonation. This is intended to facilitate the students to learn the melodic pattern of an utterance, and know the variation in the pitch level of the voice and tone and most importantly in stress and rhythm in English,

and how to convey differences of expressive meaning (e.g. surprise, anger, wariness). Two activities have been taken up to assess the importance of this sub-skill. First, the learners have been made aware of intonation. They have listened to recorded conversations, giving attention to how other speakers use intonation to express themselves. This has been followed by repeated practice. Another activity has involved students recording their own voice to listen to one's own voice because it sounds different to what we expect. Intonation is stressed/practiced with this activity.

- The applications on the sub-skill of fluency have been facilitated with the goal to make students concentrate on fluency since speaking skills involves gaining fluency in spoken interactions with others as well as practicing pronunciation. The exercises carried out under this sub-skill involve students practice speaking with a logical flow without planning or rehearsing. Activities which require students to focus on meaning in communication without immediate concern for accuracy are also carried out. A mini-dialogue in pairs has been demonstrated in class with the teacher as A and the rest of the class as B. The first line of the dialogue has been written on the board. For example, Did you go somewhere interesting last night? Students have been asked to count how many words were used (seven). Response is elicited from any student, to continue the dialogue, but with six words. For example, No. I was at home only. The dialogue has been continued with a five-word sentence; e.g. Were you at home alone? A four-word sentence has been elicited. The mini dialogue has been continued until it concluded with one word. Mini dialogues have been repeated by the participants in the same manner in pairs; A and B, since the students have enjoyed it.

- The tasks on the sub-skills of conversation and appropriateness in choice and range of words, pronunciation, grammar, and vocabulary are aimed to help the students in the context of real questions and answers as a part of conversation skills and techniques, like disruption, reiteration, clarification, information, details, turn taking, and giving in the conversation. Additionally, these sub-skills are taught with the intent to facilitate the students to reproduce the pronunciation and intonation with accuracy of vocabulary, words, structures, and pronunciation, and use language appropriately for a situation or a specific topic or task. Specific activities have been framed to demonstrate the efficacy of these sub-skills in the form of social language activities which are conducted regularly for five minutes at the beginning/end of each class ranging from who went where/when/why/how? Concurrently, questions on traditional festivals/holidays/local-national-international news/TV or videos seen/games played/school-college memories, etc., have been asked. Organizing language activities have been made a daily part of the study period classes. They have included checking attendance (who is absent? Where is Khriesanth?). Students have been also involved in organizing group activities making use of sentences like "Which group would you want to join, Khriesanth?" Participants have been also asked to do things, fetch things, give things out, etc., all aimed at natural speaking. Students have been encouraged to pose questions when they have doubts or when they need something to be repeated.

Reading Sub-Skills

Under reading, eight sub-skills have been chosen for the study.

- The applications on the sub-skills of reading for detail and identifying a referent word in a text have been facilitated with the intention to make the students to look for specific information when reading a text, information-finding skills to understand stated ideas and information in text. Additionally, the purpose of these sub-skills is to make the students understand ideas in a text which are not explicitly stated and being able to separate essential and non-essential content in text: The exercises under these sub-skills have included students being made to distinguish main idea from supporting detail. They have also read for 'drawing and doing', apart from reading lesson texts and comprehension passages, letters, faxes, and e mails for sharing details.
- Students need to apply the sub-skill of summarizing to sum up a text to give a clear picture of it in the shortest form possible, with all the key information and ideas. The students have been asked to highlight all key information and events in a text/story. Students' key points have been written on the board and discussed as a whole class. Then in groups, the participants have been asked to summarize the text/story within a word and time limit. A peer group comparison followed.
- The tasks on the sub-skill of skimming have been framed to support the students to go through a text quickly to get an overall idea of the content, without going for details or any specific information. In addition, the activities have been aimed at the purpose of rapid reading to get the main ideas/sense. Students have been asked to read a story within a stipulated time. Then the story has been read by the teacher pretty swiftly while students were to follow the lines with their eyes, not fingers, on the words. A questioning time followed in which students have to answer within a time limit.
- The tasks on note-taking/making is intended for students to understand the organization of the text and to be able to identify the main points and the supporting details, in skeleton or outline form. In the activities, the students have been asked to extract salient points for summary of specific idea with relevant and related points from text for summary. They then have learned to reduce text by rejection of redundant or irrelevant items or information.

- The assignments on local comprehension (referred also as an intensive reading skill) have been framed with the purpose to enable the students to read a piece of text closely or intensely for the purpose of extracting specific information from the text, with focus on the details of the information provided by the writer, which will generally be located in different parts of the text. Students have been asked to read few paragraphs from a given text, after which they are to ask themselves a series of specific questions in the nature of 'Whys', 'Whats', and 'Hows'. They have been told that these are only few of a host of questions to aid to the comprehension of the text. They have been then asked to go back to the passage to locate the specific information. When they focus their attention on the questions in mind and find the answers, they have been asked to connect them, considering that the specific information from the text for an understanding of a specific point is met.

- The method adopted on global understanding, comprehension, and reading for gist is intended to support students in reading from the 'whole' to the 'parts', and not vice versa, to form an overall 'picture' of the entire text. This will facilitate students to read through the text at high speed in order to identify and pick up the main idea or ideas in the text while 'filtering out' the unnecessary details. Learners have been made to read to get the overall understanding of a text. Story telling after reading has also been practiced. They have read stories to share, participate, analyze, and synthesize. They have also visited the library to read for fun and drawing or painting.

- The assignments on the sub-skills of predicting and informed guesses have been framed to make it possible for students to form a reasonable and accurate picture of what the author is trying to say from reading a few sentences, paragraphs, or pages with the ability to hop and skip through the text, omitting quite substantial portions of it without missing important information. In addition, the purpose of these sub-skills is to enhance students' predictions about what is likely to be found in the text and reading for repetition, expectation, imagination, and prediction.

- The tasks on scanning for specific piece of information in which the reader is interested have been facilitated with the intention to make the students focus on reading quickly through a text while looking for keywords to find particular details or glance through a text till an answer, key word or information is found. The main activity has involved specifying the details required in writing or otherwise. To improve speed, a time has been set which steadily reduced with the series of lessons. In the initial stages, the paragraph or line of the required information has been indicated, but not after that. The exercise has been conducted on a competition basis; in teams and individually, to encourage speed.

Writing Sub-Skills

Eight sub-skills of writing have been selected for implementation in this study.

- The assignments on planning, structuring, forming, and linking are chalked up with the purpose to make students pay attention to think actively about how texts are structured and what kinds of phrases or vocabulary are used for different purposes (e.g. introducing a topic, describing, comparing, and contrasting, writing conclusions). This is intended to improve the writing as well as reading. Students have been asked to read given texts to get ideas and to write drafts, revise structure and plan while writing.
- The tasks on describing a picture or a series of pictures have been facilitated with the objective to make the students focus on creative and guided writing with given situations and stories which could open the line to keep words flowing. A picture/series of pictures have been kept on the blackboard. Words and structures have been elicited and students have been asked to write the whole story. Each group has been given the same images or a group of pictures based on which they wrote a group story after discussion. On completion, the story has been passed around for spelling/grammar check and then pinned on the board for presentation. The same activity has been repeated with different pictures, group discussion but individual story to be written. Sequencing, spelling, and grammar have been checked by regrouping so each group has a representative of each picture. The pictures have been taken away and students have been tasked to place the story in order.
- The assignments on the sub-skill of fill in the blanks have been framed to promote students' creativity with word choice on a correct base with one word gap or more random gaps. To keep the participants interested and engaged with a game spirit, books with diverse missing words have been given to them to fill in the blanks in pairs and individually too.
- The tasks on paraphrasing have been facilitated with the goal to support students concentrate on clarity in restating written words, ideas, or thoughts of another person in appropriate word order, with good standard grammar usage. In addition, the purpose of paraphrasing is to make students know how to express a particular meaning using different grammatical forms, leading to good writing. A fun introductory activity has been conducted by splitting the students into pairs and question were asked to them such as, "What did you do after college yesterday?" or "Tell where you would like to go on vacation and why you would like to go there." Student A answered the question in three or four sentences. Next, Student B paraphrased Student A's answer. Then, the partners switched roles. This has been modeled to them twice before the actual activity. The students were also taught the difference

between paraphrasing and summarizing with different strategies and activities, like, the four R's approach of paraphrasing: reword, rearrange, realize, and recheck.

- The assignments on organizing and appropriate layout have been facilitated with the target to improve the students' awareness in recognizing the use of accurate vocabulary, idioms in writing paragraphs, topic and support cohesion and unit with cohesive devices and writing conventions for writing drafts or asking for peer correction. The students have been taught to structure a text into paragraphs and use devices such as thesis statement, to write purposefully and meaningfully. They have also written paragraphs, stories, letters, letter replies, faxes, and e mails.

- The activities on writing to a situation have been framed with the intent to develop students' writing as per the situation, as in, writing responses to everyday situations. The activity procedure has involved a group of students listening to a news item and a conversation. Another group has read a newspaper, letter, and invitation, while another group watched an experiment. Then all the three groups have written a letter (complaint/query/acceptance/refusal/thanks) based on what they listened, read, or watched. They have also practiced writing letter to a newspaper, writing report about an observation, experiment. They have written what they think/feel/know about the topic at hand.

- The tasks on proofreading have been facilitated with the intention to make it possible for the students to check a text to assess it and determine if it is needed to correct any mistakes in spelling, grammar, etc. The students have been taught how to proofread for only one kind of error at a time to maintain focus. For example, spelling errors first, then punctuation, etc., since it's easier to catch grammar errors without checking punctuation and spelling at the same time. Students have also been asked to separate the text into individual sentences to help them to read every sentence carefully, looking for grammar, punctuation, or spelling errors. Another activity has been to circle every punctuation mark to check if the punctuation is correct. Another activity has been on reading a text backwards for checking spelling. This technique has been helpful since content, punctuation, and grammar won't make any sense, with the focus entirely on the spelling of each word. Each student has written a paragraph on their favorite game/star/book/and learned to identify and proofread the specific areas of their own writing.

- The activities on summarizing have been developed with the purpose to make the participants concentrate on the ability to abridge a text with all the key points and details. The learners have been made to listen or read a text and share in their own words in the form of a summary with the main key points. Then they have extracted the main points out of a lengthy text and rewrote them in shorter forms, using full sentences.

Results

The participants have been provided activities as mentioned in the methodology of the study for a period of six months. A pre-test has been conducted before the commencement of the course and a post-test has been conducted at the end of the course. The results of the data collected are explained below.

Performance Levels in English Language Skills

Table 6.2 shows the eight performance levels followed for the study with the corresponding percentage levels achievable by the participants, where 0–30 percentage level is the lowest performance level (Below Average) and 91–100 percentage level is the highest performance level (Highly Distinguished). The students have been assessed based on the performance level as depicted in Table 6.2.

Overall Performance Level in English Language Skills before the Study Period

Figure 6.1 illustrates the average performance level of the two participant groups in English language skills before the study period. As depicted, the average performance of both the groups in English language skills stands at below average level with a total score of 28.9%. The findings of the study show that both the groups of engineering students face stumbling blocks in their overall core language skills. The study infers that the biggest possible reason could be the absence of effective and track tested methods of language teaching to garner interest and engagement in engineering students toward English language learning.

Table 6.2 Percentage and Performance Levels

Percentage Level	Performance Level
0–30	Below Average
31–40	Average
41–50	Satisfactory
51–60	Good
61–70	Very Good
71–80	Excellent
81–90	Distinguished
91–100	Highly Distinguished

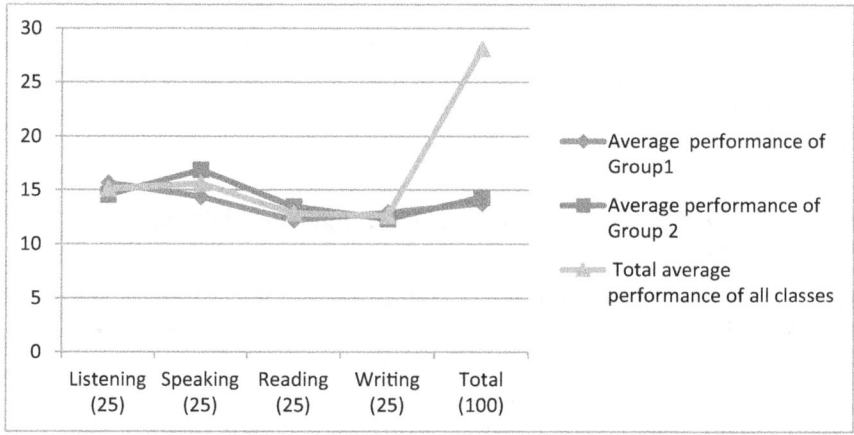

Figure 6.1 Overall Performance Levels of all Participants in English Language Skills before the Study Period.

Source: Authors.

Performance Percentage of Participants in English Language Skills in Pre-Test

Figure 6.2 reveals a below average performance in all the language skills. The test discloses a barely crossed average level of performance by the two groups of students in listening and speaking skills, while the reading and

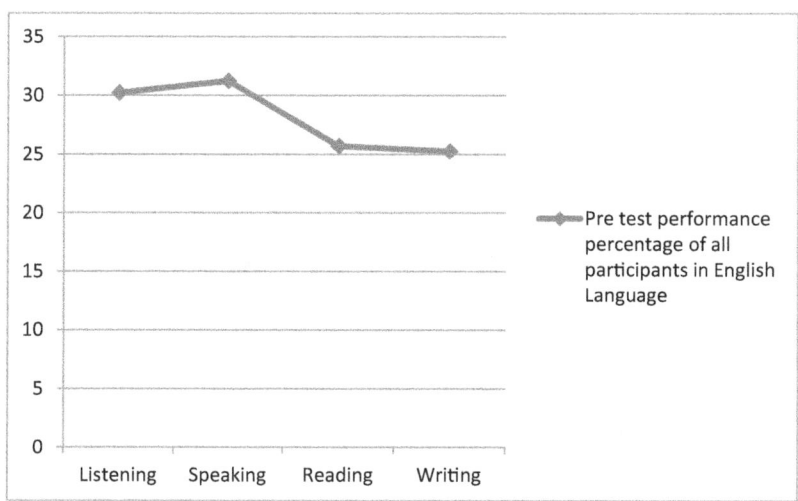

Figure 6.2 Pre-Test Performance Percentage of all Participants in English Language

Source: Authors.

writing skills are below average. The study has discovered a mixed result in the output and input language skills of the students, which construes that with appropriate teaching strategies intended at the sub-skills of integrated LSRW, the performance level of engineering students in ESL classrooms will show a different result.

Overall Performance Level in English Language Skills after the Study Period

Figure 6.3 depicts the improvement of both the groups in English language skills from below average level to satisfactory level after they are being coached for the study. The study confirms that students can enhance their core language skills if taught along with activities based on the sub-skills of English language. The facilitation of this strategy will augment students' language learning scenario in the ESL classroom.

Performance Percentage of Participants in English Language Skills in Post-Test

Figure 6.4 shows a satisfactory level of performance by both the groups of students in all the language skills. The test shows that the best performance by students after the study period is in speaking followed by listening, writing, and reading. As represented in the figure, the speaking skill got the highest score, while the reading skill is the lowest. The study findings give

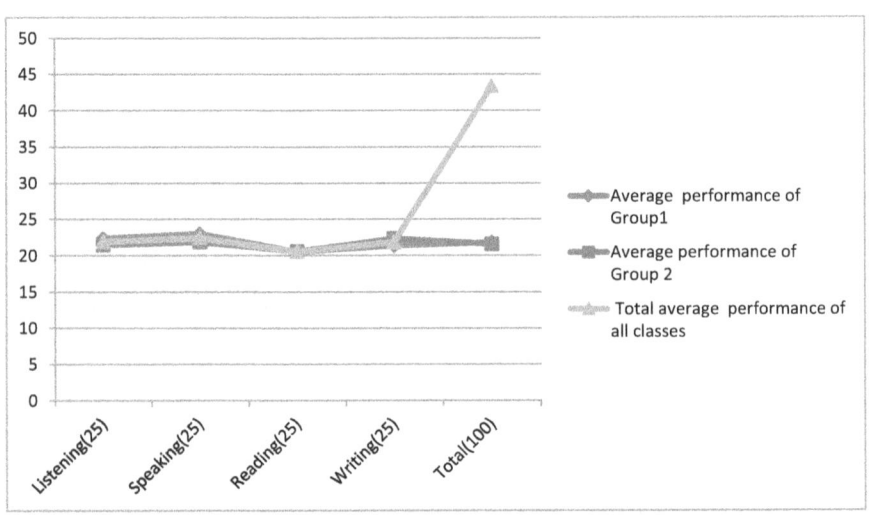

Figure 6.3 Overall Performance Levels of Participants in English Language Skills after the Study Period.

Source: Authors.

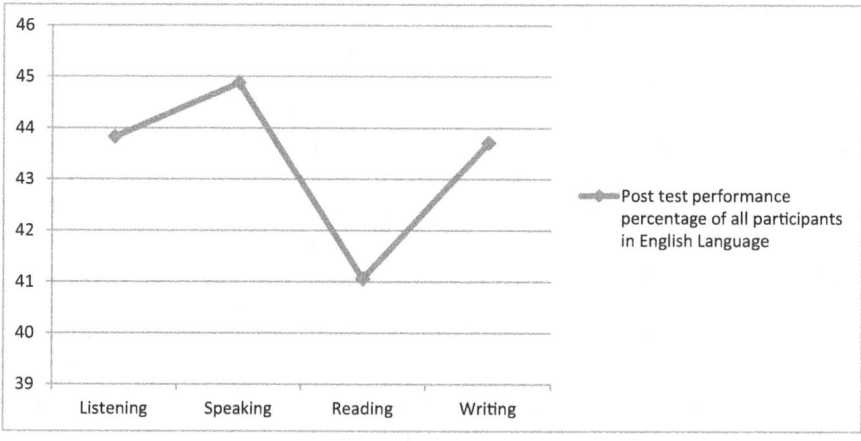

Figure 6.4 Post-Test Performance Percentage of all Participants in English Language.
Source: Authors.

evidence that a curriculum and teaching transaction design based on an integrated model of the language skills will produce a balanced improvement result in all the core skills. The findings also exhibit that the integration of activity-based language sub-skills approach will lead to experiential and engaged learning in the ESL classroom which on the whole can boost the learners' LSRW skills.

Discussion

Diverging from the studies which suggest gaining proficiency of isolated language skills in language learning (Mohan, 1986), and the ones that advocate language skills being linked to only one particular skill or the other alone (Peregoy and Boyle, 2001), the research design of this study is based on an integration of the sub-skills of the core language skills to probe the scope of teaching LSRW with engaged sub-skill activities. Language sub-skills like selective listening, analyzing, synthesizing, planning, drafting, editing, summarizing, scanning, note-making, and predicting, etc., are valid across the core language skills (Oxford, 1990).

The study results demonstrate that a premeditated plan to gain proficiency over a set of language sub-skills can lead to the improvement of LSRW skills in general. The study outcome also reveals that offering options to language learners to choose the type of sub-skills, i.e. basic, foundational, or advanced, or in a graded manner according to their level, interest, and readiness, is effective in determining the success level of students' ability and application of LSRW. The study discloses that the duration of achieving the desired outcomes for the sub-skills should deviate from being strictly stipulated. Another significant outcome of the study is the direction shown toward the advantage of being aware of the strengths and weaknesses of learners in the language sub-skills to facilitate the process of enhanced language teaching and learning.

Pervasive approaches in learning the language sub-skills have showed success in facilitating connection among the core skills altogether through the activities. Since the participants performed the least in reading and writing in the pre-test, a greater number of sub-skills (eight for both) have been purposefully selected for these two skills and the end result has showed an impressive outcome in the post-test. The research design has specifically integrated writing across the major sub-skills activities, as writing scored the lowest pre-test percentage. The experiment has resulted in the improvement of students' writing skill from 12.62% in the pre-test to 21.85% in the post test. As the study activities necessitated recurrent speaking without inhibition, the speaking skill of the students has also witnessed marked progress.

The study shows that students can improve their core language skills if activities based on the sub-skills of English language are facilitated in the ESL classroom. Mastery over the respective sub-skills of the English language helps in easing the language learning process of the learners. The study implies that the learning level of success in the learners is determined by how teachers rise to the challenge of motivating students to develop the required language skills and keep them engaged. Efficient and stimulating techniques, however, minor, are needed to keep learners interested, energized, and on task. The focus is aimed at a range of techniques and approaches in sustaining interest and motivation, while ensuring that learning takes place. Learners' responses to examples of motivating activities particularly were given special focus since they give out whether learning is happening or not as expected.

Conclusion

The purpose of the present study is to examine the role of sub-skills in LSRW. After rigorous analyses of the theories, language proficiency tests, and course books, a list of conclusive sub-skills has been prepared. The sub-skills have been implemented in the classroom through activities and techniques which acted as a motivation to the students to participate enthusiastically in the activities. The results have revealed that the core skills are not entirely separate entities and that all the skills consist of similar and different sub-skills. This study offers a list of operational sub-skills which could prove helpful for teachers and students alike to not only improve the core language skills but also to design or take tests and prepare course materials. The pre- and post-tests of the core skills have been designed with the listed sub-skills with well-designated activities and strategies. The importance of observing the students while participating in the skill activities or answering test questions on the measured sub-skills has been perceived in the course of the study. The pre- and post-test scores analysis, students' think-aloud practices, and the interviews with the participants explain the dimensionality of the LSRW sub-skills. An improved performance of the learners in the sub-skills of LSRW has been noticed. The findings exhibit that the integration of activity-based sub-skills teaching and learning in the ESL classroom enhances the core LSRW skills of the students.

References

Alderson, J. C. (2005). *Diagnosing foreign language proficiency: The interface between learning and assessment*. London, UK: Continuum.

Bloomfield, L. (1933). *Language*. New York: Henry Holt and Company.

Brown, H. D. (2001). *Teaching by principles: An interactive approach to language pedagogy* (2nd ed.). New York: Pearson Education.

El Menoufy, A. (1997). "Speaking. The Neglected Skill". *New directions in speaking. Proceedings of the fourth EFL skills conference*. Under the auspices of the Center of Adult and Continuing education the American University in Cairo. pp: 9–18, Egypt.

Faerch, C., & Phillipson, R. (1984). *Learner language and language learning*. Clevedon: Multilingual Matters Limited. (Vol. 14).

Field, J. (2008). *Listening in the language classroom*. UK: Cambridge University Press.

Goh, C. C., & Aryadoust, V. (2014). Examining the notion of listening subskill divisibility and its implications for second language listening. *International Journal of Listening*. 29(3). Doi: 10.1080/10904018.2014.936119

Goodman, K. S. (1967). Reading: A psycholinguistic guessing game. *Journal of the Reading Specialist, 6*(4), 126–135. Doi: 10.1080/19388076709556976

Harmer, J. (1997). *The practice of english language teaching*. London: Longman.

Hoshino, Y. (2010). The categorical facilitation effects on L2 vocabulary learning in a classroom setting. *RELC Journal, 41*(3), 301–312. Doi: 10.1177/0033688210380558

Kennedy, S., & Trofimovich, P. (2010). Language awareness and second language pronunciation: A classroom study. *Language Awareness, 19*(3),171–185. Doi: 10.1080/09658416.2010.486439

Lundsteen, S. W. (1979). *Listening-its impact on reading and the other language arts*. Urbana, IL: National Council of Teachers of English. ERIC Clearinghouse on the Teaching of English.

Mohan, B. (1986). *Language and content*. Reading, MA: Addison Wesley Publishing Company.

Oprandy, R. (1994). Listening/speaking in second and foreign language teaching. *System, 22*(2), 153–175.

Oxford, R. (1990). *Language learning strategies: What every teacher should know*. Boston, MA: Heinle & Heinle.

Oxford, R. (1996). *Language learning strategies around the world: Cross-cultural perspectives*. Manoa: Second Language Teaching & Curriculum Centre, University of Hawaii Press.

Oxford, R. (2001). *Integrated skills in the ESL/EFL classroom*. ERIC Digest. Washington, DC: ERIC Clearing house on Languages and Linguistics.

Peregoy, S. F., & Boyle, O. F. (2001). *Reading, writing, & learning in ESL: A resource book for K-12 teachers*. New York: Addison Wesley Longman, Inc.

Richards, J. C., & Rodgers, T. S. (2009). *Approaches and methods in language teaching* (2nd ed.). Cambridge: Cambridge University Press.

Richards J. C. & Reppen, R. (2014). Towards a pedagogy of grammar instruction, *RELC Journal, 45*(1), 5–25. Doi: 10.1177/0033688214522622.

Rivers, W. (1981). *Teaching foreign language skills* (2nd ed.). Chicago: University of Chicago Press.

Sanchez, M. A. A. (2000). An approach to the integration of skills in English teaching. *Didactica, 12*, 21–41. Retrieved from https://revistas.ucm.es/index.php/DIDA/article/viewfile/DIDAD000110021A/19603.

Su, Y. C. (2007). Students' changing views and integrated skills approach in Taiwan's EFL college classes. *Asia Pacific Education Review, 8*, 27–40.

Vandergrift, L. (1999). Facilitating second language listening comprehension: Acquiring successful strategies. *ELT Journal, 53*(3), 168–176. Doi: 10.1093/elt/53.3.168.

Part IV

English for Specific Purposes (ESP)

Chapter 7

Value Addition to the Communication Course

S. Mekala and C. Harishree

The development of new technologies and challenges in the workplace in the era of globalization has transformed our communicative purposes. The contemporary digital era has taken the communication to multidimensional levels in both physical and virtual landscapes. Pearson group (n.d.) has defined communication as "a social process in which information is exchanged in order to convey meaning and achieve desired outcomes". Apart from the need for communicating in their academic tasks every day, the students must be exposed to the challenges of effectively and efficiently communicating in the workplace.

In the engineering sector, communication skill has become more prominent and vital for the students of engineering in recent years. Students of engineering need communication skills for different purposes to communicate their thoughts and ideas, negotiate and guide the team members and consumers, analyse information, etc., to execute at different levels of oral and written communication. Public speaking, documenting, presenting, instructing, and guiding a team are some of the important communication skills required in the engineering workplace. Students of engineering in India are skilful in their technical subject but most of the students lack effective communication skills, which in turn affects their employment opportunity. To overcome this problem, many institutes, and organizations like AICTE emphasize on developing communication skills of the students to meet the workplace challenges. There have been researches conducted on analysing the communication skills gap of the students of engineering. The findings of few researches in reputed institutes have proposed English for Specific Purpose (ESP) on communication skill to enhance students' skills set. Indian educational department has propagated communication as part of the curriculum to improve the students' communication skills. Yet, the students' communication skills and proficiency have not raised to the expectations of the industry. It is opined by Donnell et al. (2011) as, "Although engineering department have worked hard at improving the communication skills of their students a large percentage of industry managers consider the communication skills of engineering graduates to be weak". Consequently, it necessitates to reorient the communication skills courses

DOI: 10.4324/9781003268529-11

offered to the students of engineering in India. Correspondingly, ESP courses on communication skills for the students of engineering are researched by scholars and teachers in India. These researches concentrate on developing or integrating an ESP course syllabus in the engineering curriculum based on the needs of the students. But there are scanty studies conducted on integrating project-based learning methodology in the English classroom of the students of engineering to develop their communication skills. Hence, this chapter will focus on adding value to the current communication skills course of the students of engineering in India through project-based learning. It could help the teachers to develop the students' communication skills to meet the industrial requirements, redefining their current teaching-learning process.

Need for the Study

Due to rapid changes in the workplace demands of the 21st Century, there is a need to upgrade the syllabus of the students to equip their skills set. Communication skills is the most essential skill set required in today's global workplace. In concord with this, the current English syllabus of the engineering curriculum concentrates more on communication skills, although listening, speaking, reading, and writing skills remain the primary concern of the language classroom. ESP plays a vital role in meeting the language needs of the students and helps the teachers to concentrate on developing the professional expertise integrating specific disciplinary knowledge and requisite skills set. The focal point of communication needs analysis of the students of engineering has transformed ESP as an efficient course to develop the students' communication skills. The integration of ESP courses on communication skills into engineering curriculum has improved the students' skills set to an extent, but the employers are not completely satisfied with the students' language proficiency. The communicative course of the students of engineering in many institutes in India has been more of a teacher-centred focus rather than student-centred orientation in the prescribed syllabus. This has led to the lacunae in the practical application of the acquired skills set, resulting in constrained communication of the students in real-life situations. Hence, it is necessary to implement activities in the English communication course that involves real-time problems related to the workplace communication lapses. Further, it provides opportunities for the students to analyse the situational needs of the communication process and overcome the lacunae that leads to the communication skills gap in their workplace environment. In this line, the current chapter analyses the administration and implementation of case studies on communication lapses to the first-semester English course of the NIT-T students and discusses the effectiveness of the project-based learning in improving the communication skills of the students.

Developing ESP Course and Classroom Activities

According to Johns and Dudely-Evans (1991), the components of ESP course design comprise needs assessment and discourse analysis. As needs analysis of the students plays crucial role in designing the ESP courses, the ESP practitioners and material designers have been constantly improving the needs analysis techniques. Needs analysis comprises different steps and methods encompassing necessary data collection on learners through various possible means. Information of learners has been collected under two categories namely, demographic details of the learners like their social and cultural background, and learners' learning preferences and their perceptions on the need (Nunan, 1982). Based on the collected data, the content of the course will be developed by policy makers and educationalists. Consequently, discourse analysis is a primary factor in ESP course development examining the purpose and usage of language to a specific context. It is commonly associated with analysing both the written and spoken language with respect to social context. It helps in developing the students' communication skills to meet the demands of their future workplace environments.

Further, ESP course development involves strategies and methods to incorporate the content in the English classroom. ESP courses prescribe learner-centred course design. So, the methods and pedagogical strategies will be framed with the learners' involvement in their learning process. In concord with this, Klimova (2015) has proposed task-based language learning (TBLL) method in ESP course development. TBLL focuses on assigning different classroom activities and tasks to improve the students' communication skills along with developing their subject knowledge. Some of the TBLL activities in ESP courses for enhancing communication skills are simulation games, report analysis, problem-solving activities, debates, vocabulary building exercises, etc. Lavrysh (2016) has proposed role games, interviews, conferences, group discussion, project creation, etc., as some of the activities that can be incorporated in the ESP courses to improve the students' communication skills. Further, the author has proposed communicative approach to the English language learning, cooperative learning, and problem-based strategies to enhance the oral communicative skills of the students. Further, project-based learning activities will help the students in understanding the subject with in-depth analysis of the real-time problems of the workplace situations. Hence, integrating project-based learning activities in the ESP course design can add value to the course in developing the students' communication skills.

Collaborative Learning

Collaborative learning provides opportunities for the students to communicate with their peers while working on particular project or tasks. Gol and

Nafalski (2007) have asserted, "the notion of collaborative learning alludes to learning within groups that have been formed for the specific purpose of achieving set of educational goals". Further, collaborative learning instigates discussions with peers developing inter-group and intra-group communication among the students. Concordantly, Kirschner (2001) explains, "Collaborative learning supports the use of effective discursive learning methods (make explicit, discuss, reason, and reflect, convince) while allowing for the acquisition of essential social and communication skills". Correspondingly, collaborative learning method can promote the students' communication skills set. Smith and MacGregor have suggested five premises that makes collaborative learning successful namely,

- Learning is an active process
- Learning depends on rich context
- Learners are diverse
- Learning required social environments
- Learning has affective and subjective dimensions

(as cited in Ahmad & Yunus, 2019)

In consonance with this, project-based learning is an active process that depends on rich context, which involves students with diverse language ability. It also provides different social environments according to the students' future workplace requirements comprising both affective and subjective dimensions. The project-based learning methodology can be used as an effective collaborative learning approach to develop the students' communication skills in the teaching-learning process of the English course of the students of engineering.

Project-Based Learning

Project-based learning is an active learning process, in which students understand their subject concepts through evaluating the provided information that requires self-directed learning skills set. Putri and Hidayat (2019) have defined project-based learning as, "a learning system that provides independent learning to students, constructive investigations, goal orientation, negotiation, collaboration, communication and reflection in learning situations related to the real world". Project-based learning methodology provides opportunities for the students to work in the development of a creative project that involves complex tasks requiring collaborative approach at each stage say analysing the solution, implementing the ideas, solving the problems, and making decisions with the learners. Major (2001) has stated that project-based learning "helps students develop advanced cognitive abilities such as creative thinking, problem-solving and communication skills" (as cited in Awang & Daud, 2015). Moreover, this methodology is considered to be more appropriate for engineering courses and widely used in the language teaching during recent years. Congruently, Morimoto (2016) has stated problem-based learning methodology "has been extended to engineering programs and has been as effective in improving problem-solving and communication skills".

According to National Academy of Science and Engineering, the industrial development of the 21st Century has posed unique challenges to new engineers (Alves et al. 2017). It necessitates the different pedagogical approaches and methods to implement the industrial requisite knowledge in the ELT classroom. Correspondingly, project-based learning is instrumental in developing a collaborative learning environment that can help the teachers to instigate the students' interest and make them involved in their learning process, by executing the project-based activities with their peers. Arboleya and Las-Heras (2014) have stated that the project-based learning methodologies focus on the communication activities that can help students to develop their teamwork and communication skills and to make them adept in the competitive edge of the global workforce.

Project- vs Problem-Based Learning

Problem-based learning is a student-centred approach similar to project-based learning and it involves the students to solve challenging real-life problems as a classroom task. Project-based learning focuses on assigning the students with creative project output and problem-based learning aims to provide an opportunity for the students to understand the problem and provide solutions. In both learning methodologies, students require self-directed learning skills and especially in problem-based learning, the learners have to depend on their own knowledge and there is a need to identify requisite information to solve the real-life situation problems. Problem-based activities also involve group project works and students need to communicate their thought process and solve the specific problem assigned to their group. The problem-based learning encompasses the stages say framing list of hypotheses, searching, and discussing requisite information, developing problem statement, analysing the gathered information, and evaluating their findings with their group members and presenting the most effective solution in front of their classmates. The main purpose of the problem-based learning is not finding the solutions to the problem but understanding the cause of the problem. So, problem-based learning evokes critical thinking skills and improves effective communication skills of the learners. In accordance with this, when the students are made to work on communication lapses in the workplace environment, it can ascertain them to overcome the lacuna in real-life application of the global platform, which in turn satiates the desired outcome of the communication course of the engineering programme.

Case Study

Case study is analysing the real-life situations at the workplace for class-room discussion between students and teachers. It helps the students to improve their skills set and knowledge through discussion-based learning environment. Case study method has been primarily used in non-linguistic fields like medicine and law and later introduced at Harvard Business School in 1910 to help the students analyse hypothetic business problems (Nae, 2019). The method is later adopted by other disciplines due to its effectiveness in teaching methodology. Case study method is often used in both project-based learning and problem-based learning environment. While administering case study methodology, students are required to answer open-ended questions and provide different potential solutions to real-life workplace problems. It is a flexible method as instructor can chose a structure of implementation of subject knowledge and skills in their classroom environment. Case study has procured significant position in language teaching methodology during recent years in the ESP context. In this case study methodology, the students can be acquainted with the workplace scenarios and be prepared to face the challenges in their future workplace, in the process of enhancing their language skills. It necessitates the students to read the case, speak and discuss the situation, listen to different perspectives, and document the inference or solution and archive the content. Students can improve their content generation, vocabulary, and grammar pertaining to their domain in this pedagogy.

The case study provides opportunity for the students to apply their creativity and professional knowledge in analysing the real-life situations. It also promotes the decision-making skills of the students through logical argument during case study discussion. Case studies are enumerated in different categories say scientific research case studies, practical case studies, educational case studies, etc. Practical case studies will reflect the problems that occurred in real-life situation at detailed description that could help the students to understand their particular work field. Scientific research case studies deal with the cases on researches and educational case studies concentrate on academic oriented cases. The case study helps in developing various skills set of the students like critical thinking, problem-solving, organization skills, collaborative skills, and communication skills. This method involves different stages like individual preparing the information on the particular case, discussing his/her information with the members of the group, and presenting in the classroom in front of their peers and facilitator. On the whole, it will develop the communication skills of the students, inclusive of the nuances of delivery and techniques of presentation.

There are different approaches to implement case study method in the English communication course. According to Kelch and Malupa-Kim (2014), the instructors have to provide the materials related to the case study before the discussion in the classroom that could help the students to be familiar with the particulars and prepare for the classroom discussion. Further, the authors have stated that

> instructor/facilitator may guide discussion of the topic, theory, or issue before transitioning to group work on the case. Groups then discuss the case, presenting their opinions, clarifying the issues, proposing solutions, reaching opinions, or whatever the particular task calls for.
>
> (p. 14)

In this line, the current chapter discusses on administering case studies on the broader theme of communication lapses as a group project work and make the students of engineering to discuss the subthemes with their team members and present their inferences on the particular case in a collaborative learning environment.

Methodology

This chapter focuses on the project-based learning methodology administered to the students of the first-semester English course of B. Tech programme at National Institute of Technology – Tiruchirappalli (NIT-T). After the introductory sessions on case studies, the students were divided into groups and a group leader was assigned. Contemplating on the idea of project-based learning and exploring the possibilities of communication, the broader theme was suggested to the students. Further, students were asked to select two case studies under subthemes, say social, political, historical, economical, ethical, medical, human-centric lapses. The students were given a week's time to register this group project with their facilitator.

The rationale of assigning this major theme of 'Communication Lapses', is to make the students comprehend the nuances of the communication process by working on various lapses, so that they could avoid such lacunae in their prospective careers. They have been promoted to think ahead and take a holistic perspective of the communication process. In this line, the students have been introduced to a case study, the difference between a story and a case study, and the nuances of case study presentation and they have also been briefed on the ground reality that lapses occur starting from our day-to-day conversations to the topmost level executive of a corporate. The thin demarcation line between barrier and lapse has been explicated to the students, as any problem initiates with a small barrier resulting in great communication lapses, leading to loss of human lives, loss of tie-up between concerns, productivity loss and even war between two countries in certain cases.

The students have been asked to present their analyses on the selected case study and submit a comprehensive document of their work on those case studies. Students have been also asked to provide rationale for selecting particular case study to make them comprehend the significance of the real-life crisis occurred due to communication lapses. The chapter will discuss the case studies on Charkhi Dadri Mid-Air Collision, Tenerife Airport disaster, Beas River Tragedy, Chernobyl Nuclear Disaster, Downfall of Nokia, General Motors Cobalt Ignition Switch. The students' project presentation and inferences on these case studies was well received by their classmates. The students have been facilitated to view the case studies in different perspectives, which in turn contributed to interactions and discussions in the teaching-learning process.

Implementation

Students presented their case studies on their subthemes on communication lapses, by analysing the situation of the crisis, cause of the problem, communication barriers that had occurred in the cases, and provided the alternative solutions that could have avoided the crisis.

Case Studies on Aviation Disaster

The first case study is on 'Charkhi Dadri Mid-air collision'. It is about a flight crash of Saudi Arabian Airline and Kazakhstan Airline over the village of Charkha Dadri in 1996 killing 349 people. The cause of the accidents has been identified as miscalculation of altitude measurement by the pilot, weather condition, deviation from original flight schedule, and inadequate airport infrastructure. But, the ultimate cause of this accident has been reported as the communication problems between radio operator and pilot of the Kazakhstan plane. The students have reported the rationale for selecting this case as pilot's lack of English language knowledge, delayed communication, misconception of communication due to differing social and cultural background. Students have analysed these reasons and worked on solutions to overcome such barriers in the workplace. They have also evaluated the solution taken on the account of avoiding such accidents in future. They have also reported their solutions like mandating the knowledge of English language for the higher crew authorities like pilot and flight captain to avoid miscommunication, and also suggested that radio operator should be aware of the measuring equipment of flights before giving the altitude directions to avoid miscalculation. Consequently, the students have analysed the crisis situation and the negligence of the pilot on the warning from the controller tower due to lack of language skills. Further, the students have understood the lack of effective communication that can lead to massive death trolls as indicated in this case study.

Similarly, the case study on 'Tenerife Airport Disaster' deals with the flight crash between American (Pan Am 1736) airline and the Netherlands (KLM 4805) airline in 1977. This case study has been selected for project work, as it has served as an example for future accident prevention frameworks and investigating such accident in aviation industry. Later this accident has been made as a mandatory study in pilot training on Crew Resource Development course. It is the most notable case study, as the accident occurred majorly on communication lapses between the flight crew and operating tower. The students have identified misconception, categorical thinking, non-standard usage of words, noise in communication channel like visual noise and psychological noise, and organizational barriers as reasons for communication lapses. The students worked on this case study has engaged in analysing the final conversation between flight captain and tower operator. The students have analysed the part of conversation to identify the usage of words that led misconceptualized on the particular context. For example, the pilot was enquired about the ATC (Air Traffic Control) clearance but he has been received the instructions to be followed for takeoff. The communication has not reached the pilot properly due to technical glitches in radio communication that leads to miscommunication. Similarly, the categorical thinking of the pilot of not taking the flight engineer's questioning about another flight in the runaway and reluctance of the engineer to question the pilot's approach to takeoff were also considered to be reasons for the communication lapses. The students have identified that the usage of diction plays a vital role at the workplace communication in this case study. The students have learned the significance of employing appropriate diction and judicious usage of phrases and clauses to avoid such crisis in their future workplace.

Case Studies on Power Plant Disaster

The Chernobyl nuclear disaster, the notable case study based on massive man-made disaster, is presented by students for analysing the communication lapses. Though the case has been used in different disciplines for various aspects of investigation, the students have worked on the communication issues in this project. The disaster occurred at 1986 and its impact is still in the soils of the city of Pripyat transformed it into an uninhabitable land. The cause of the disaster has been identified as miscommunication between the engineers. The students have indicated the work pressure of the vice chief engineer and multiple transfer stations that have led to this massive disaster. Antoly Dyatlov is the vice chief engineer in charge of nuclear power station and he forced his junior Aleksander Akimove to run the test on reactor without providing him proper guidelines on safety precautions. He was running short of time to submit his report and he urged his junior to conduct the test with his authority. Congruently, the students have analysed the cause of the communication problem as higher authority's improper instructions and junior's hesitance in obtaining proper guidelines from higher authority. For instance, if Dyatlov had communicated better with his junior on safety measures instead of urging him to run the test and if Akimove had conveyed his reluctance more strongly with risk factors on running the test, the disaster could have been avoided. Further, they have also identified the barriers of communication as categorical thinking, varied perceptions, fear of superiors, and job insecurity. Consequently, the students have also inferred that minor communication lapses can result in major crisis in the history of mankind.

Further, students have worked on '2014 Beas River Tragedy' that focuses on drowning 14 second-year engineering students and one tour operator. This case study is based on an accident that occurred at the Beas River in Mandi district of Himachal Pradesh, India. It happened because the Larji Hydroelectric Power Plant has opened its floodgates without prior intimation. The students who have been taking photos on the river bed drowned in the flood within few seconds of opening the dam. This tragedy has occurred due to the negligence of the higher authorities and lack of following proper protocols for opening the dam. Students have inferred the communication negligence of government officials that has led to this disaster. They have done extensive research on the reports of the tragedy from sources say interviews, newspaper articles, the internet write-ups on the accident, and government press releases. The locals have claimed that they have blown whistle to warn the students, but it was in vain due to the distance between the locals and students on the riverbed. The authorities have claimed that they have warned prior to the open of dam by siren, but it has also not reached the students due to the sound of the river flow. Hence, the students have concluded that inappropriate medium of communication has led to loss of lives in this tragedy. Besides, the students have learned that any communication should be channelled through proper medium to avoid crisis.

Case Studies on Corporate Disaster

Students have presented on prominent case studies on communication lapses that have led to corporate disaster and one such notable case study is the 'Downfall of Nokia'. The students have gathered information on the integral flow of communication within an organization through this case study. The students have analysed the various levels of communication and its impact on the workplace situation in this case study. As future engineers, they have been able to acquire the workplace environment and organizational structures pertaining to corporate sectors from such case studies. In the 21st Century digital world of work, the future graduates always required to hone skills to thrive in the global competition. The students need to understand that the communication skills play vital role in the global competition and the workforce have to communicate at different levels of an organizational structure. The students have presented the reputations of higher authorities and common demotions and firings have made the organization lack to maintain proper hierarchy for communication. Similarly, lack of acknowledgement of product's lower efficiency by top managers have also led to the downfall of the company. Further, the middle managers did not hold strong position for making decisions on critical situations. This made the workforce perplexed in executing the ideas and perceptions of top managers that affected the production. It has led to the gradual decline of the quality of the product. Moreover, the students have identified both organizational and interpersonal communication barriers in this case study. The students have suggested that Nokia's downfall could be prevented, if the personnel have approached their anxiety of failure in positive manner and worked in a healthier communicative situation. Further, the students could provide insights on overcoming communication lapses in critical situations at workplace.

The next case study on 'General Motors Cobalt Ignition Switch Crisis' has marked a serious language usage issue that has led to the death of 13 people and 32 car crashes. It is about the inefficient decision making of General Motors authorities that led to a catastrophic turn of events. The study deals with the Chevrolet Cobalt car's ignition switch that caused the failure of electrical system and airbags of the car. The usage of words in the report submitted by engineer has made General Motors to consider the critical issue as a minor issue. The issue has been ignored by the managers of General Motors, as the employees were instructed to use the words like "issue, condition or matter instead of problem" and "does not perform to design instead of defect". It has led to mention the major safety problem as customer convenience problem in the report and the ineffective communication of critical safety information on ignition switch of the Cobalt. Moreover, General Motors experts paid no attention to the issue considering it to be least important to reconsider the usage of words. In this case study, the students have analysed the usage of diction and phrases and the vital role of communication in the workplace. The students have identified that this communication lapse could have been avoided in face-to-face communication with the use of appropriate non-verbal cues like facial expressions, gestures, etc. The students have also indicated the importance of the content in PowerPoint presentations. The employee has stated the deaths of five people in a bulletin of the appendix during his presentation, which the managers and other authorities neglected. The students have stated that bulletin has reduced the clarity of the fatalities related to this issue and instead of bulletins, the problem could have been stated clearly with a plain text and simple communication. In this case study presentation, the students have learned the significance of written and spoken communication and the occurrence of communication lapses at all levels of an organization and its serious impact on the organizational crisis.

Implications

The case study method could be implemented even in a mixed ability group of large number of students in the classroom. Communication and presentation skills of the students have been developed in the execution of this case study project work. For instance, after grouping the students, they have discussed on the rationale of selecting a particular case study with their group members. The students have shared responsibility among their group to complete the project. The documentation of their work and their presentation of the case study in the presence of their classmates have helped in overcoming their anxiety and reluctance of communicating in English. The analysis of case study on communication lapses has provided them an insight of workplace communication roles and also helped them to work on avoiding such communication lapses. The students have been assessed on the basis of presentation of their case studies and their role in the final documentation. It helped the students to identify their personal communication lapses and work on developing their communication skills.

The case study method has promoted the students' active involvement in the teaching and learning process. It facilitated the critical thinking ability by analysing every aspect of the case and identifying the role played by communication lapses on the crisis. The students have demonstrated their analytical skills and their knowledge on the role of communication in different real-life situations. The students have also developed their communication skills in accomplishing this project venture. Additionally, the students have developed – collaborative skills, creativity skills, and critical thinking skills – the most essential skills set of the 21st Century workplace situations. The case study method has provided opportunity for the students to understand the organization structure and working nature of different fields. The explicit presentation and comprehensive analysis of the case studies have conveyed that the students are able to comprehend the intricacies of communication. Most of the students have confessed in their presentations that they would avoid such lapses in their prospective career.

Conclusion

In the context of globalization, every student is expected to be competent in English language and proficient in their communication skills. Mastery in communication skills has gained a lot of significance and value and it has become a prerequisite of an incumbent in the corporate sector. Effective English language communication skills help students to enhance their self-confidence and self-esteem levels. The students of science and technology should go beyond communication skills to perceive and comprehend a holistic perspective of communication and use it in their real-life situations. Consequently, the communicative course of the students of engineering needs to be improved with possible methods of pedagogy. Apart from the need for communicating in their academic tasks every day, the students must be exposed to the challenges of effectively and efficiently communicating in the workplace. In this line, the chapter has explored the case study method as an effective pedagogical method implemented to improve the communication skills of the students of engineering in the 'English for Communication course' of the engineering curriculum. Hence, it is recommended to implement such project-based learning methods to add value to the English communication courses of the engineering programme.

References

Ahmad, M. A. K. B. A., & Yunus, M. M. (2019). A collaborative learning intervention module to improve speaking fluency. *International Journal of Scientific & Technology Research, 8*(12), 1834–1838.

Alves, A. C., Leao, C. P., Moreira, F., & Teixeira, S. (2017). Project-based learning and its effects on freshman social skills in an engineering program. In M. Otero-Mateo & A. Pastor-Fernandez (Eds.), *Human capital and competences in project management.* United Kingdom: IntechOpen Limited. Doi: 10.5772/intechopen.72054

Arboleya, A., & Las-Heras, F. (2014). Improving independent learning and communication skills of students in last year of engineering degrees through the use of project-based learning methodologies. *IEEE, 21,* 1–7. Doi: 10.1109/TAEE.2014.6900149

Awang, H., & Daud, Z. (2015). Improving a communication skill through the learning approach towards the environment of engineering classroom. *Procedia–Social and Behavioral Sciences, 195,* 480–486.

Donnell, J. A., Aller, B. M., Alley, M., & Kedrowicz, A. A. (2011). *AC 2011-1503: Why industry says that engineering graduates have poor communication skills: What the literature says.* Washington: American Society for Engineering Education.

Gol, O., & Nafalski, A. (2007). Collaborative learning in engineering education. *Global Journal of Engineering Education, 11*(2), 173–180.

Johns, A. M., & Dudely-Evans, T. (1991). English for specific purposes: International in scope, specific in purpose. *TESOL Quarterly, 25*(2), 297–314.

Kelch, K., & Malupa-Kim, M. (2014). Implementing case studies in language teacher education and professional development. *ORTESOL Journal, 31,* 10–18.

Kirschner, P. A. (2001). Using integrated electronic environments for collaborative teaching/learning. *Research Dialogue in Learning and Instruction, 2,* 1–9.

Klimova, B. F. (2015). Developing ESP study materials for engineering students. *Proceedings of 2015 IEEE global engineering education conference (EDUCON),* (pp. 52–57). Estonia: Tallinn University of Technology.

Lavrysh, Y. (2016). Soft skills acquisition through ESP classes at technical university. *The Journal of Teaching English for Specific and Academic Purposes, 4*(3), 517–525.

Morimoto, C. (2016). Improvement of IT students' communication skills using project based learning. *Proceedings of the 8th international conference on computer supported education (CSEDU 2016),* (Vol. 2, pp. 147–152). Rome: Italy.

Nae, N. (2019). Teaching English with the case method – A tentative approach. *Euromentor Journal, 1,* 25–38.

Nunan, David. (1982). *Language teaching methodology: a textbook for teachers.* Englewood Cliffs: Prentice-Hall.

Pearson Group. (n.d.) Communication: Executive summary for educator. Retrieved from https://www.pearson.com/content/dam/one-dot-com/one-dot-com/global/Files/efficacy-and-research/skills-for-today/Communication-ExecSum-Educators.pdf

Putri, S. U., & Hidayat, S. (2019). The effectiveness of project-based learning on students' communication skills in science. *Journal of Physics: Conference Series, 1318,* 1–7. Doi: 10.1088/1742-6596/1318/1/012006

Chapter 8

Poetry in the Engineering Curriculum

S. P. Dhanavel and S. Kumaran

In the past four or five decades, functional uses of language have gained importance. With the rise of English for Specific Purposes (ESP) and communicative language teaching (CLT), English is taught for its functional and communicative value in various streams of education, including technical education. The syllabus for functional and technical English is designed and aimed at developing communication skills in students of engineering. However, the teaching and testing of different kinds of courses on English are usually directed towards passing students in English to achieve a target set by the college or university boards of studies. When students face placement interviews and/or attempt to get jobs through them, the recruiters generally complain about the poor communication skills of students as majority of them are visibly unable to communicate competently and confidently in English. There may be several reasons for the glaring mismatch between what is taught and what is learnt. One of the possible major reasons is the dissociation between the affective and cognitive dimension of students when they learn English. If students are able to pass the subject in the examination, what prevents them from passing the test of everyday communication in English? In order to address this issue, use of poetry is suggested for the learning of English in the context of technical education for students of engineering and technology. The functional and technical communicative syllabus can be explored through poetry as an effective resource for emotionally linking the subject, the skill, and students so that they will learn to communicate through English in real-life situations, including placement interviews.

In today's globalised and technocratic world, the scope of poetry in the teaching and learning of English language at the tertiary level, especially technical education is being neglected. In this context, this chapter explores the possibility of teaching some of the teaching points from the technical English syllabus, including listening, speaking, reading, and writing skills, through poetry.

DOI: 10.4324/9781003268529-12

AICTE's Document on Technical Education

All India Council for Technical Education (AICTE) is the national statutory body vested with powers to maintain the quality of technical education in India. Recently, it has come out with a number of initiatives to improve the quality of technical education in the country. While it has diversified the humanities education for engineers in tune with the Washington Accord to include history, sociology, psychology, economics, management, it has also suggested addition of elective courses from literature, including the course on Elements of Literature. However, a bunch of courses in Technical English, Professional English, Functional English, Basic English, Communicative English, Communication Skills, Communication Skills Laboratory, etc., continue to dominate the Technical Education scenario across the country, although they have not completely addressed the issue of communicative competence of engineering students in real-life situations like placement interviews.

Language and Poetry

Fundamentally, language is metaphorical and thus poetic. Whether it is the language of common people or that of engineering professionals, the process of communication is one of transforming experience into language symbolically. As language and poetry are highly charged with emotion and thus connotation and suggestiveness, poetic communication is enriched with meaning. As Collie and Slater (1987) observe, "Reading poetry enables the learner to experience the power of language outside the straight jacket of more standard written sentence structure and lexis. In the classroom, using poetry can lead naturally on to freer, creative written expression".

English and Communication through Poetry

English language can best be learned through poetry from common nursery rhymes to mind-blowing experimental poems. The learner's and the poet's attempt are always aimed at conveying meaning exactly and effectively. A thorough training for technical students in reading, understanding, appreciating poems in English, including poems in English translation can be greatly beneficial to students. In fact, poems can effectively be used to teach the four language skills of listening, speaking, reading, and writing. While discussing Cleanth Brooks' notion of the heresy of paraphrase, C T Indra (1995) observes that paraphrase is a useful tool not only for learning a language but also for developing students' linguistic sensibility. She says, "Language exercise may be constructed to test the students' sensitivity to the original language of the poem and that of the paraphrase" (21).

Critical Views on Poetry and Technical Education

Developing the Interpersonal Communication Skills (IPCS) of students at college level is emphasised by the academia and the industry to bridge the gap between academic skills and employability skills. Among the qualities sought by the employers from their employees, IPCS are considered the most important. Poetry is basically a rich source of emotional and interactional diversity. Therefore, a variety of emotions embodied in poetry in English from various countries can be a valuable resource in developing the IPCS of students (Dhanavel & Ramaraju, 2015). Behind this illusion, of course, is the technical writer, shaping and synthesising the inchoate stuff of experience into reports designed to inform and enlighten an identified audience, and thus engaging in a fundamentally poetic process. Science and technology are far more poetic than some of their practitioners and, ironically, most humanists seem to realise (Rutter, 1985).

Poetry for Engineers

Engineers by definition have to solve problems creatively. Exposing engineering students to poetry can enhance the creativity of students through an understanding of the figurative nature of language. Although engineers have to achieve precision in their communication, paradoxically it is poetry that can have a long-lasting impact on students in conveying meanings unambiguously. Actually, Millan (1996) has shown that poetry and engineering can be integrated meaningfully through a joint project of Engineering and Humanities for developing both linguistic and creative skills of students.

Sample Poems and Teaching Points

Poems of all kinds can be used for engineering students. However, the samples used in this chapter are largely famous sonnets from English literature.

William Shakespeare's "Shall I compare thee to a summer's day?" begins with a rhetorical question "Shall I compare thee to a summer's day?" which holds scope for the would-be engineers who ought to rely on their power of communication to achieve a successful career. Rhetorical question is one of the means to persuade the listener to accept the deliberations of the speaker, and hence, introducing engineering students to this literary device enables them to outshine others by mastering the art of effective communication. Further, it would help them add variety to their speech and to be different from their counterparts.

Rhetorical questions and metaphors are the vital part of effective communication and the engineering students learn the nuances of raising questions for the sake of convincing others and winning their confidence. They will understand the prospective nature of using metaphors in their day-to-day communication and will try to incorporate them to develop their argument. When the students get exposed to the way Shakespeare mesmerises readers by his dexterous use of metaphor, they too learn the art of using language a bit different from its ordinary purpose.

Through Shakespeare's poem, students have a wonderful chance to learn the art of using metaphor. In fact, in this poem, the poet does not expect an answer rather raises a question so as to convince the reader with his thoughts. After much labour, he finds a suitable metaphor to ascertain the greatness of his beloved. Though it seems that the poet wants to compare his friend to "a summer's day," he discredits the choice by calling his friend more lovely and temperate as he is not transient like the summer day. Summer days are known for their varying manifestations such as heavy wind, scorching sun, and fleeting weather. On the other hand, he claims that the charm of his friend "shall not fade" and such a handsome young man needs more than an ordinary metaphor. He realises that art alone has the power to immortalise the sublime quality of his friend and substitutes his poem as an alternate metaphor that captures the perennial beauty of his friend.

John Donne's "Death be not proud" is a good example for the engineering students to master various techniques of communication such as irony, rhetorical question, personification, comparison, assertion, and command. The poet personifies "Death" by considering it as a persona. Moreover, the poet begins the poem by addressing a person and such learning is quite useful for the students to attribute human touch to inanimate things and abstract entities. Such a practice would help the students to convert any kind of situations to their advantage by charging them with feelings and emotions. The poet develops his argument in such a way that the readers' mood moves with the thought of the poet and it helps students to develop their argument logically and forcefully. The students can very well notice how the poet grasps the attention of the readers by the opening line itself thus: "Death, be not proud, though some have called thee/Mighty and dreadful, for thou art not so;" and they learn how to write a viable topic sentence. They can also notice how the argument is developed systematically by extending the metaphor till the final line. The assertion nullifies the power of death over humans as the poet accentuates the fact that death is not the end, rather it is a beginning of the journey of one's soul into another region. There is also an instance of comparing the sleep induced by death with that of sleep gained by "poppy" and it enables the students to work on analogy. Students have the chance to learn a novel method of communication by assuming the persona as an intended listener, as the poet has assumed "Death" as his addressee. Such a gesture would enable the students to propose their thoughts without the restriction of their intended listeners. Further, students have a scope to achieve the power to command as the poet does by saying "Death, thou shalt die."

William Wordsworth's "Upon Westminster Bridge" is a reliable example to have a different perception towards an event or a situation. As engineering students are the backbone of industries, they ought to understand how

Wordsworth has viewed river Thames in the context of industrialisation. The poet does not subscribe to the common view that industrialisation has contaminated the pristine nature of London city. Rather, the poet describes the beauty of the city in the early morning before the office hours of industries and highlights how nature retains its charm even in adverse circumstances. In a sense, engineering students can find in it a convenient medium to meditate the possibilities to ensure methods that safeguard the purity of their physical environment. The poem also allows students to be unique in their perception of things and to express themselves without succumbing to collective opinion against their conviction. Students can notice the capitalisation of the initial letter of the word "City" and it demonstrates how the whole place is considered a unit that holds all people and things. They can also examine how the City is filled with "Ships, towers, domes, theatres, and temples" rather than with people and it proves they are counterparts to humans. Further, engineering students can boast of the poet's use of "Bridge" as an image in the poem. It is the engineers who are the creators of bridges and by their creation they find a medium to connect humans and nature. The image of "Bridge" shows how humans cherish their bond with nature using the platform created by them. On the whole, the students would realise that the poet has achieved a profound impact due to his power of description. This power relies on the use of appropriate images such as the "smokeless air," "sun," "valley, rock, or hill," "houses seem asleep," and "mighty heart is lying still!" and their realisation would help them to exercise the power of description in their career.

John Keats's "Bright Star" relies on imagery and symbol to express the conviction of the poet and the exposure of engineering students to these techniques would facilitate their ability of logical reasoning. The "star" is used as a symbol that validates the wish of the poet as he wants to be "stedfast" like the star. In fact, the symbol permeates all through the poem and allows the poet to use different tones depending upon the progression of his thoughts. In the beginning of the poem, the tone of the poet is complacent as he wants to imbibe the quality of being "stedfast" like the bright star. However, he contradicts his own statement and uses ironic tone when he declares he would not be passive like the star. Ultimately, the tone becomes emotional at the end of the poem as the poet wants to feel the warmth of his beloved forever transcending the limitations imposed on nature on humans and thereby resorting again to the steadfastness of the star. "Sleepless Eremite," "the moving waters," "earth's human shores," "soft-fallen mask of snow," "the mountains and the moors" signify the power of appropriate images to the engineering students to shift their tone casually without losing the intended effect on the listeners. The students gain the skill to develop their thoughts logically by using symbols and images that are relevant to the situation they want to address and they also become proficient in influencing the minds of the listeners by adjusting their tone according to the mood of their listeners.

Gerard Manley Hopkins's "Thou art indeed just, Lord" presents a common situation which every engineering student experiences at times and hence an apt choice for learning. They can very well relate the dilemma of the poet with that of themselves and pose questions to the creator. The poet is baffled with the happenings of the world and is confused as good people suffer whereas people of unscrupulous nature enjoy all the gifts. Like every human, the poet is unable to suppress his troubled thoughts and poses many questions to God asking Him to reveal the decorum of happenings. This is the exact situation which students would feel when they are clueless of the order of things. However, the rational questions posed by the poet to God educate them about developing their argument by providing a justification. At the same time, the students would learn how to maintain their composure though involved in interrogation like the poet who addresses God as "O thou my friend." It reveals how one should never resort to foul language just because of being impatient and exemplifies humility as a safe guide towards a moral path. The poem educates the students to compare their progress with that of the objects of nature and to justify their achievement. The poet does it by comparing his performance with that of nature thus:

Sir, life upon thy cause. See, banks and brakes/Now, leavèd how thick! lacèd they are again/With fretty chervil, look, and fresh wind shakes/ Them; birds build – but not I build; no, but strain, "and he realises whatever he has tried to build crumbles whereas the birds build nests that withstand heavy wind". Though the poem is full of interrogation, it ends with a promising prayer "Mine, O thou lord of life, send my roots rain".

John Donne's "No Man is an Island" brings out the values of humanity and interrelationships among the engineering students as the poem deals with the unity of all lives. In addition to interpersonal skills, students can also learn how to use conditional clauses to express themselves in an effective manner. The poem reveals that humans cannot live in isolation as: "Every man is a piece of the continent, /A part of the main." The use of "Anaphora" is a common device that involves repetition of certain words and the effective use of "If clauses" makes the students think of their bonding with other people and to cherish the relationship thus: "If a clod be washed away by the sea, /Europe is the less. /As well as if a promontory were. /As well as if a manor of thy friend's/Or of thine own were:" Moreover, the students have the chance to understand that the death of their fellow beings is the loss of their own power and it also reminds them of their own doom: "And therefore never send to know for whom the bell tolls; /It tolls for thee." Further, the beauty of the poem lies in its creation of mood and tone achieved by the poet by his generalisation of particular events. Such a skill ought to be learned by the students as it excavates their subdued feelings to cherish the universal brotherhood. The poet achieved the effect by his metaphor of comparing the loss of humans with that of the loss of certain portions of continents and motivates the students to preserve the human bonding wisely.

P. B. Shelley's "Ozymandias" familiarises the engineering students with techniques such as relative clause, reported speech, and direct speech and encourage them to use these tools in their regular conversation. In fact, the poem is written in a conversational tone as the poet is found in conversation with a stranger who reveals the dilapidated monument of a late emperor who conquered the whole world thus: "Who said: 'Two vast and trunkless legs of stone/Stand in the desert. Near them, on the sand, /Half sunk, a shattered visage lies, whose frown, /And wrinkled lip, and sneer of cold command." Though it seems that the conversation takes place between the poet and the traveller, the students can feel that the poet is speaking to them directly in the pretext of addressing an invisible listener. Further, the poet's attempt at making the Emperor speak directly to the readers enables the students to feel one with him thus: "My name is Ozymandias, king of kings: /Look on my works, ye Mighty, and despair!" The poem allows the students to learn the impact of metaphors as the poet brings out the impermanence of pomp and power by his subtle use of the monument as a metaphor. It makes the students comprehend the fact that time ravages everything and the only way one can overtake the impact of time is by using Art as a resort. Hence, this poem makes the engineering students to achieve both the understanding of human life and the use of conversational techniques to engage their listeners in the process of expressing their thoughts and feelings.

William Blake's "A Poison Tree" helps the engineering students to learn complex and compound sentences in addition to anger management. It also proves how the students can create tremendous impact on the listeners even by the use of simple language. The poem examines two ways of anger management which would be very helpful for the students to adopt in their work place. They understand how anger loses its power when one reveals it to the person concerned whereas it develops immensely if it is hidden and kept a secret thus: "I was angry with my friend: /I told my wrath; my wrath did end. /I was angry with my foe: /I told it not, my wrath did grow." The students also learn the use of metaphor being employed in the poem. The garden represents the growth of anger which the poet waters it with his tears and fears and which bears apples that take the life of his enemy by alluring him to consume it. When the students are aware of the deceptive nature of anger, they would easily manage themselves in their work place keeping themselves away from harming others and from losing their peace of mind. They understand that the best way to manage anger is in expression rather than suppression and they safeguard themselves from its pitfalls. Further, the whole poem is written in complex and compound sentences such as "And I watered it in fears, /Night and morning with my tears; /And I sunned it with smiles, /And with soft deceitful wiles." and they help the students in expressing the progress of their thoughts. This is one of the suitable poems for mastering emotional intelligence.

Robert Frost's "The Road Not Taken" exemplifies the development of logical reasoning through compound sentences like "Two roads diverged in a yellow wood, /And sorry I could not travel both" and progressive verbs such as "Yet knowing how way leads on to way" among the engineering students. In fact, the repetition of "I" in the above quote signifies the responsibility thrust on the person who makes choices and it cautions him to be careful while exercising his freedom in selecting choices in his life and career. Students are exposed to a dilemma regarding choices in life and are cautioned to select choices after careful scrutiny of available factors. Moreover, they learn how to use compound sentences while reading the first stanza of the poem and realise the equal opportunities available in choices through the open ending of the poem thus "And that has made all the difference." Moreover, the poet uses future tense "I shall be telling this with a sigh" to evaluate the influence of the past on the present and concludes the poem with a statement which becomes a universal statement. The students have the occasion to realise the necessity to frame statements of universal significance and resort to compound structures as adopted by the poet. John Milton's "On His Blindness" creates an opportunity for students to involve in self-analysis like the poet who examines his life after becoming blind. He regrets that the creative talent granted to him by God is still unused by him and longs to utilise his talent to produce a sublime work. Students will get motivated to analyse their strengths like the poet and will try to achieve something great in their lifetime. They will also be made to realise that it would be inappropriate to waste their talent by keeping it away from being fruitful. However, they will also understand that patience would be a reliable guide to achieve greatness, as God has many people at his command to carry out tasks ordained by him. Moreover, the engineering students have a chance to share their thoughts through self-analysis, which is different from direct expression of the same. They have the opportunity to learn passive construction to ascertain their idea with force thus: "When I consider how my light is spent." Further, they can learn complex syntax that distinguishes itself from ordinary use of language. In fact, they have the opportunity to learn the use of figurative language that permeates the entire poem and which also aids in shifting of mood in the final line: "They also serve who only stand and wait." The poet's mixed use of complex and compound sentences can educate the engineering students to order words thus: "That murmur, soon replies: God doth not need /Either man's work or his own gifts: who best/Bear his mild yoke, they serve him best. His state/Is kingly; thousands at his bidding speed/And post o'er land and ocean without rest:"

As the poem "Night of the Scorpion" written by Nissim Ezekiel is known to most of the readers, it can very well be chosen to examine how LSRW skills can be developed among the engineering students.

Listening Skills

Listening helps enormously in oral fluency development. A teacher can enhance the listening skills of the students through pre-listening, while-listening, and post-listening activities and she has to do some homework to achieve her target. A teacher has to give a short lecture on the importance of listening skills and the listening to the lecture may act as a pre-listening activity as it prepares the groundwork for the understanding of the poem. When the teacher reads the actual poem, the students should be encouraged to identify the pronunciation of complicated words and to mark various types of intonation used. The teacher can check the understanding level of the students through post-listening activities. She can ask the students to spell out some of the complicated words and make them answer questions related to the poem. Further, the teacher can make the students fill up the blanks created by her. At this stage, the interaction is between the students and the teacher and is not among the students themselves. The interaction among the students should be reserved for the development of speaking skills.

Speaking Skills

A teacher can give a short lecture on vowels, consonants, diphthongs, into-nation, stress, and other para-linguistic features and make the students gain a preliminary knowledge about them. Further, she can initiate conversation among the students by asking some questions related to the poem. As the teacher knows the theme, she can ask questions such as:

1. Who is the persona of the poem?
2. Do you have any personal experience like that of the persona?
3. What is the theme of the poem?
4. What do you think of village life?
5. How do you consider the belief of the villagers?
6. How do you understand the word 'love'?
7. How is mother's love superior to all other forms of love?
8. What is meant by the following lines?
   "With candles and with lanterns throwing giant scorpion shadows on the mud-baked walls."
9. Who is referred to as he in "May he sit still, they said"?

When the students finish answering the questions posed by the teacher, they can be made to involve in group discussion (GD) and role play. For GD, the topics such as

1. Science versus Tradition
2. Superstition versus Reason

3.   Ancient versus Modern

can be given and the students have to be encouraged to discuss the topics with interest and involvement. Further, the students can be assigned some roles to perform. They can play the role of the persona of the poem, the sceptic father, the poor mother, the peasants, and even the scorpion.

Reading Skills

A teacher can show the picture of a scorpion and make the students gain preliminary knowledge about the poem as a part of pre-reading activity. In addition, she can use materials related to the theme of the poem to throw better light.

> The teachers' job, prior to the reading of any poem, is to create the kind of mental landscape that will ease the students into the poem...The trigger should be something outside of the poem that can readily touch the lives of students, and then be linked into the poem. The trigger can consist of pictures, a film strip, a quotation, an anecdote, or any other device that seems suitable.
>
> (Hess, 2003)

As a while-reading activity, the teacher can ask students to note down the expressions that proclaim the sarcastic tone of the poet or they may be made to identify the words that express the sufferings of the mother. Further, the teacher can give choices for select words and ask students to identify the right choice. She can also encourage students to analyse the impact of chosen words on the readers. As a post-reading activity, the teacher can make students rearrange a line or a few lines.

> When the poem has become very much our own, students enjoy working closer with the language. They take turns re-reading the poem to each other in pairs. I give them the openings of lines, and they complete them from memory and later check their results with the poem. They write definitions of single words, and later match their own and classmates' definitions to the words.
>
> (Hess, 2003)

The post-reading activity helps in the development of writing skills too.

Writing Skills

A teacher can help students exercise their knowledge of grammar and sentence structures through various activities. She has to choose appropriate grammar items that are evident in the poem and expose students to them or she can make them write a paragraph or essay on select topics related to the poem. Some of the grammar exercises that can be taught through the poem "Night of the Scorpion" include:

Words	Outcome
was stung, was not found	Voice
steady rain, diabolic tail	Adjective
had driven, risked	Tenses
I, he, it	Pronoun
Him, they	Direct/Indirect
Parting with	Phrases
with, for, by, of, on	Prepositions
mud-baked	Compounding
Mother's blood	Possessive
More candles, more lanterns	Degrees of comparison
Sceptic, incantation, hybrid	Vocabulary
the bitten, a match	Articles

The teacher can also nurture the creative writing of students by making them write their own poems.

In this present-day world, people lead a compartmentalised life and they do not generally feel affinity with other humans and non-human others. In fact, some people view this world as a hostile and an indifferent place where they can find no reason for living. Moreover, they are frustrated with their lives and go to the extent of taking their own lives in order to escape from day-to-day drudgery. In this regard, Kahlil Gibran's *The Prophet* can be used to enlighten the engineering students on topics like love, marriage, children, eating and drinking, work, joy and sorrow, laws, self-knowledge, teaching, good and evil, prayer, friendship, religion, and death.

Some of the factors that prove fatal to the sanctity of humanity include exploitative capitalism, corruption, terrorism, and humans' selfish acts. Engineering students would learn from the prophet to come out of their narrow boundaries and to realise their oneness with the universe.

Self-awareness reveals the importance of understanding the basic nature of oneself and it helps everyone to realise their abilities to cope up with the hard realities of life. Only through self-awareness one can develop other skills. The prophet educates people about the necessity of sacrificing oneself for the sake of others. He believes that people should not judge the deserving status of the borrowers rather should give everything as a tree benefits both the good and the bad. Moreover, he says that rather than giving things one should surrender oneself for the betterment of the world and the sacrifice would turn the world as a paradise on earth: "You often say, 'I would give, but only to the deserving.' / The trees in your orchard say not so, nor the flocks in your pasture. / They give that they may live, for to withhold is to perish."

Coping with stress and emotions refers to identifying the situations that make people lose their balance of mind and which disturb their inner peace. It is only by deciphering their origin and root cause, people can come out of their brutal impact. As the progress of mankind leads only to the problematic situations which have no solution to address the shortcomings, the prophet urges humans to realise the causes that aggravate the imbalance of mind. Humans often fall a victim of stress and emotion as they fail to comprehend the real nature of good and evil. The prophet points out that one has to cope with stress and emotions as they arise due to the whims and fancies of humans and they can keep them pure by abstaining from unethical deeds. It is only by following the path of dharma one can be free from the clutches of the swinging mind: "You are good when you are one with yourself. /Yet when you are not one with yourself you are not evil. /For a divided house is not a den of thieves; it is only a divided house."

Conclusion

Evidently, poems can teach engineering students the best of communication skills for performing their job successfully. If used properly and creatively by mapping poems with certain teaching points from the Technical English syllabus, they can connect English language with students strongly, thereby helping students draw on the rich resources of language throughout their life. However, care should be taken while choosing poems for the class.

> When designing materials to use with a poem, teachers should firstly analyse what is unusual or distinctive about the language in the poem. The materials or tasks for students should be devised around these unusual features, since this will increase both their understanding of the poem and their knowledge of the language in general.
> (Lazar, 1993)

References

Collie, J., & Slater, S. (1987). *Literature in the language classroom: A resource book of ideas and activities.* New Delhi: Cambridge University Press.

Dhanavel, S. P., & Ramaraju, S. (2015). Developing the interpersonal communication skills of college students through poetry: A classroom study. *IUP Journal of English Studies, 10*(2), 54–63.

Hess, N. (2003). Real anguage through poetry: A formula for meaning making. *ELT Journal, 57*(1), 19–25.

Indra, C. T. (1995). *Teaching poetry at the advanced level.* Chennai: TR Publications.

Lazar G. (1993). *Literature and language teaching: A guide for teachers and trainers.* New Delhi: Cambridge University Press.

Millan, H. L. (1996). Poetry in engineering education. *Journal of Engineering Education, 85*(2), 157–161.

Rutter, R. (1985). Poetry, imagination, and technical writing. *College English, 47*(7), 698–712.

Chapter 9

Incorporating Thinking Skills in the Engineering Curriculum

S. Mekala and C. Harishree

The 21st Century has brought remarkable achievements in technological innovations and development. The development of Industry 4.0 and invention of artificial intelligence have taken the role of physical employees in many workplaces. Students of the 21st Century are competing with the technological advancements of new digital era to achieve in their career. Furthermore, the COVID-19 situation has transformed the physical workplace to digital platforms where the computer and the internet play a major role. It necessitates the workforce to augment the skills set that are unique and most essential to meet the ever-changing demands of the future workplace. It is a challenge for higher educational institutions and curriculum developers to overcome the shortcomings of the students' skills set and prepare for their career prospects. So, there has been emphasis on soft skills development of the 21st Century students. Lavy and Yadin (2013) have stated "many synonyms are used in the industry for non-technical skills such as 'soft skills', people skills', 'emotional skills', employability skills', etc." (p. 1). Further the authors have stated "among non-technical capabilities, one may find strong analytical and critical thinking skills to thrive in a competitive global environment". Correspondingly, soft skills encompass thinking skills and it is important for the students to get hired in the world of work. According to Levy and Murnane, "there has been a steady increase in the need for workers to perform non-routine tasks that require complex communication and expert thinking skills" (as cited in Soule, 2020). Thinking skills are related to mental activities that comprise mental processing of the information/content, connecting different perspectives of the information, making decisions based on this mental process of information, and creating novel ideas. In this line, the chapter concentrates on developing critical and analytical thinking, and creative and lateral thinking skills of the students of engineering.

Thinking Skills

Aristotle has selected rationality "the capacity to think" as the defining attribute of man. Thinking can be perceived as a polymorphous concept which gives the meaning of 'reasoning', 'reflecting', 'pondering', 'believing'.

DOI: 10.4324/9781003268529-13

Thinking is involved even in performing an activity and while we practise an activity, we will first observe, describe, and account for that activity before performing it. This psychological inquiry is a never-ending process like one's pulse or heartbeat. We never stop thinking in our waking moments and even indulge in snatches of thought during sleep. So, this thinking potential in students has to be tapped and cultivated to achieve higher level performance in their working places.

Bloom et al. (1956) have classified analysis, evaluation, and creating as the higher order thinking skills of educational domain. Likewise, Butterworth and Thwaites (2013) have classified thinking abilities say reasoning, creative thinking, and reflection as higher thinking skills. Hall (2008) has stated creativity skill involves "designing, constructing, planning, producing, inventing, devising and making". Accordingly, the process of critical thinking is defined as gathering information and resources as a part of analysis, trying to answer questions formed to solve problems by imagining different possibilities which is a part of synthesis and finally evaluating the results from synthesis (Tran, 2013). In this line, both critical thinking and creative thinking involve processing the information gathered by experience and produce novel ideas.

Moseley et al. have stated, "thinking as human activity involves cognition (knowing), affect (feeling) and conation (wanting and willing)" (as cited in Nageswari, Ravikumar & Jayamani, 2016, p. 688). Some theorists believe that thinking skills are inborn and it cannot be taught in classroom. But it is a mental process where mind formulates thoughts by personal experience to describe various problems and situations. Through classroom activities, teachers can develop this mental process and enhance the students to think creatively and critically. De Bono has developed a lateral thinking model called 'lateral thinking' and 'parallel thinking' with the insights from his analysis on the mechanism of the mind (as cited in Madhavaiah & Ram, 2016). Second language acquisition requires the students to have different approaches to the language usage according to the context. Correspondingly, lateral thinking will help the students to break the regular learning patterns and help them accommodate to the different aspects of language usage.

Thinking Skills for Effective Communication

Communication skills are the most requisite skills set of the present digital era and it encompass thinking skills apart from the basic classification of LSRW skills. These thinking skills enable the learners to be fluent in their content generation process. But communication skills of the engineering graduates are considered to be inadequate by industries (Donnell et al., 2011). Communication skills are essential for engineers to write memorandum, reports, employee manuals, letters, faxes, contracts, advertisements, brochures, or news releases, cover letter and other paper works (Sheth, 2015). Effective communication avoids conflict in understanding information and helps to maintain good client relationship. It is very essential for team development and decision-making process, unifying at times the different perspectives of peers. Hence, it is inferred that engineering students are required to acquire effective communication skills. Consequently, communication skills are correlated with thinking skills, as both the initiator and respondent have to exercise their cognitive domain and compose logical cohesive content from their repertoire. In order to communicate effectively, the students are required to possess vivid thought, astute listening and conveying the messages in the optimum approach and create the desired impact. Congruently, developing the thinking skills of the students of engineering can eventually improve their overall communication skills and help them succeed in their career ladder. Moreover, current engineering industry requires its employees to demonstrate "accuracy and fidelity, clearness of actions, practical thinking, attentiveness and systematization. They all come from the ability to think critically to provide a reasonable argument and show a professionalism" (Ivleva, 2016, p. 1). Further, the author has proposed to introduce activities based on critical thinking in all the courses from humanities to profession-oriented courses of the students of engineering. Consequently, the project-based learning pedagogy can be promoted in English course, where the learners have the space to think and function in multiple knowledge domains. In this line, the chapter focuses on the implementation of project-based activity by assigning case studies to foster thinking skills of the students of engineering.

Learner-Centred Curriculum

Curriculum plays a crucial role in developing the knowledge and skills set of the students. The difference between traditional curriculum development and learner-centred curriculum development is the involvement of the learners in framing their learning needs, style, and content along with the teachers. Traditional curriculum can be a hindrance for the teachers to improve the skills set of the students within the given time frame. In traditional curriculum, there are pre-fixed content, strategies, and evaluation steps prescribed to be implemented in the classroom. But in the learner-centred curriculum, students are engaged in making decisions on the content selection based on their learning progress. The teachers can adapt new methods and approaches according to the proficiency and learning progress of the students. Hence, employing the learner-centred curriculum in the English course will be efficacious in developing the thinking skills of the students of engineering.

There are few challenges in adapting the learner-centred curriculum like students' interests and involvement in the learning tasks. In this curriculum, the teachers can only be a facilitator making the students take the responsibility over their learning. It will be difficult for the students to adapt the learner-centred curriculum at the beginning as they have been trained in the teacher-centred institutions in their schooling. But, the learner-centred curriculum provides different options with special activities that help teachers to make the students learn beyond their curriculum. Further, students will be actively involved in their assessment and evaluation process and they can get holistic approach to their learning experience which can help them in their life-long learning. The main principles considered in this curriculum development is to provide freedom for teachers and students to develop the skills and knowledge naturally, to transform teachers into facilitators instead of lecturer, to instigate students' motive to complete the learning tasks, and to meet the needs of the learners to succeed in their future life and career. Congruently, to foster the thinking skills of the students, the teachers should encourage students to express innovative lateral and creative opinions in the classroom. Further, they can allocate tasks or activities in the language learning classroom based on the thinking skills to enhance students' thinking ability.

David Nunan (1988) has proposed the key elements of the curriculum development model as planning procedures that involve collecting information about the learners and grouping them as learner group, selecting the content and gradation, methodology like selecting the learning materials and classroom activities, continuous monitoring of the students' progress, assessment, and evaluation. In brief, learner-centred curriculum development involves pre-course planning procedures (needs analysis), content planning, methodology (strategies, techniques, methods, tasks, activities, etc.), material selection, assessment, and evaluation. In consonance with this, the chapter has analysed the need for project-based learning, and proposes content planning and methodology of administering case studies to develop the thinking skills of the students of engineering.

Needs Analysis

Needs analysis is the process of finding the needs of the target learners for the present and future use of the knowledge. Richterich and Chancerel are the principal proponents of the use of needs analysis (Nunan, 1988). Richards (1984) has suggested that needs analysis serves three main purposes like, "it provides educational institutes and course designers need information about the learners' capacity and their need of the skills or language to meet the future workplace demands". Needs analysis in English for Specific Purposes course refers to the course design. In the process of needs analysis,

> the language and skills that the learner will use in their target professional or vocational workplace or in their study areas are identified and considered in relation to the present state of knowledge of the learners, their perceptions of their needs and the practical possibilities and constraints of the teaching context.
>
> (Basturkmen, 2010, p. 19)

There are two types of need to be analysed before designing a course: objective and subjective. Objective needs are derived from the learners' information, their use of language in real-life situation, and their language proficiency. Subjective needs comprise cognitive and affective needs of the learner at particular learning environment. So, information about affective factors and cognitive factors like personality, confidence, attitude, individual learning strategies, etc., are analysed on subjective needs analysis. The process of needs involves target situation analysis, discourse analysis, present situation analysis, learner factor analysis, and teaching context analysis.

The needs analysis plays a vital role in the course design process. The Indian engineering curriculum developers need to have holistic analysis of the students' needs while designing the course, so as to enable the students thrive successfully in their future workplace. Scanty studies have been conducted on the need for developing thinking skills of the students of engineering. The students of engineering in India have been prescribed communication skills as a part of their engineering programme. But, the communication skills gap still persists and results in the deficit of students' employment opportunities. It may be due to the communication course being not on par with the industrial needs and not complying with the elements that enhance students' thinking skills. In order to bridge this skills gap, the course designers have to analyse the lacunae in current course and students' language proficiency, and should revamp the syllabus time to time according to the changing needs of the industry.

Need for Thinking Skills in the Engineering Curriculum

The engineers need creative skill to achieve the goals of their organization, as creativity plays a significant role in the success of an organization in globally competitive economy. Creativity is mistakenly taken for arts but it deals with all domains (Piirto, 2011, p. 1). Creativity in engineers helps to work in high-tech fields by thinking novel solutions and adapting to the changing requirements of workplace. It helps to be open-minded and invent new ideas and to synthesize all possible solutions. "Many of the fastest-growing jobs and emerging industries rely on workers' creative capacity- the ability to think unconventionally, question the herd, imagine new scenarios and producing astonishing work" (Partnership for 21st Century Skills, 2008). In the workplace situation, it is essential for the engineers to think in new and unconventional aspect, identify the problem, diagnose alternative solutions for the problems, and create innovative ideas. The product or response produced by engineers will be judged based on its novelty and its appropriate usage and 'heuristic rather than algorithmic' (Adams, Ripper, Zander, & Mullins 2010). The real-world application of technical skill set of the engineers will require this pertinent creative skill.

Critical thinking plays a crucial role to maintain leadership qualities, to build team work and to save time in decision-making process in the workplace situations. The engineers in this digital era are expected to take different roles in their workplace and will be forced to make decisions on their own at crucial situations. It is essential for engineers to explore the shortcomings of the final product and to evaluate the quality of final product. Engineers should have objective perspectives in observing the workplace activity and evolve conclusions. It will aid them to communicate their findings to both the common and the technical audience. Consequently, it is necessary for the students of engineering to acquire critical thinking skills. Though the role of creativity and critical thinking is vital in the workplace for engineers, it is often ignored in the engineering education (Cropley, 2015). Hence, creative thinking and critical thinking should be fostered among the students of engineering to place themselves in good stead in the world of work. As a consequence, this chapter recommends to integrate thinking skills activities in the teaching-learning process of the English course. So, it can help the students of engineering to communicate effectively according to the situational needs of the workplace.

Content Selection

To improve the thinking skills of the students, the teachers can select the content involving students' cognitive domain, i.e., to analyse, synthesize, evaluate, and communicate the mentally processed information. Teachers and curriculum designers may have variety of sources to plan a course content. But in learner-centred curriculum, the content should be devised by consulting and negotiating with the learners after considering their needs (Nunan, 1988). The prime consideration is the level of learners from low proficiency to advanced proficiency, as it forms a pre-requisite in selecting the appropriate course book, learning materials, learning activities, and tasks. The objectives of the course play a crucial role in content selection of the appropriate modules in the syllabus. The main objective of this chapter is to develop the thinking skills of the students of engineering. So, the content selected to teach in their English course should enable the students' thinking skills say previewing current materials available and updating the classroom discussion topics, adding activities to current course syllabi according to the classroom environment, organizing the course materials by building upon the differing perspectives of the learners, etc. It is also expected from the teachers to deliver the content using appropriate methods and tools considering the course objectives and number of students. Based on the classroom strength, the teachers should modify their content delivery mode. In concord with this, methodology signifies a cardinal role in developing the students' skills set.

Project-Based Learning

English language teaching adopts different methodologies according to the discipline of the target audience. Project-based learning is a learner-centred pedagogy that involves students in hands-on training activities to acquire the requisite knowledge and to develop the skills set of the students. This activity-based learning plays a decisive role in developing the thinking skills of the students. The conventional classroom will not be sufficient to develop the student's language and skills for their workplace. The activity-based learning has advantages of adapting different learning styles according to the students' interest and involvement in the task. The activities can be inquiry-based learning, exploring students' knowledge level and thinking ability. Students will be more engaged in the learning process as it motivates students to identify multiple perceptions, imaginative, and intuitive solutions for various real-life situational problems. Hansraj (2017) has indicated, "Activity-based learning involves reading, writing, discussion, practical activities and engagement in solving problems, analysis, synthesis and evaluation. Active learning is also defined as any strategy 'that involves student in doing things and thinking about things they are doing" (p. 4434). In accordance with this, implying classroom activities in the English language classroom can help the students to analyse their thinking process and improve their thinking ability. The main purpose of this activity-based teaching is the direct involvement of the students in the learning process and take responsibility of their learning styles and strategies.

In concord with this, project-based learning involves classroom activities stimulating students' involvement to find solutions for authenticate intriguing questions or problems. Besides, project-based language learning is a welcome alternative to the dreary study and practice of specific language skills. They add variety to the curriculum, which reawakens students' involvement in the course and improves their motivation (Chastain, 1976). In fact, "Project based learning redefines the boundaries of the classroom. No longer are students confined to learning within four walls" (Simkins et al., 2002). It also leads to learner autonomy, as students take more responsibility in the learning process by discussing it with their peers and a sense of satisfaction and achievement are always seen at the final stages of the projects. According to Eisner, we should recognize that students have many interconnected parts, and when we teach, design curriculum and assess, we need to recognize the entire student community. "Attention to such complex matters will not simplify our tasks as teachers, but it will bring education closer to the heart of what really matters" (Eisner, 2005). Contemplating on the idea of project-based learning, it is felt very much appropriate to make them work on thinking skills.

Woodrow (2018) has stated that problem-based learning and case-study approach as commonly used project-based language teaching. Further, the author states that both approaches are student-centred and focus on finding a solution for a problem that requires the students to collaborate and consider the consequences of the solution they suggest at the end of the task. Moreover, Woodrow (2018) has suggested that case-study approach is "more complex and may comprise multiple sources of information and documentation" and has possibility for number of solutions. Hence, it can be perceived that the case study involves lot of thinking process which can help the students to develop their thinking skills.

This chapter focuses on the desideratum of assigning case studies on thinking skills to the students of engineering, which in turn enhances the horizons of students' thinking skill. It demonstrates the project-based learning framework that has been administered in the second semester of English course to the students pursuing B. Tech. at National Institute of Technology – Tiruchirappalli (NIT-T).

Methodology

The students were introduced to case studies and the methodology of analysing it. Further, the students were divided into small groups comprising eight to nine students in a group and a group leader was chosen by their team members. They were given a week's time for selecting any two case studies pertaining to creative, lateral, critical, and analytical thinking. They were also briefed on working with four choices namely, lateral or creative or critical or analytical thinking case studies and present the case studies in the class as a group project work. But lateral thinking case study was made mandatory, as the facilitator intended to motivate the leaners with more discussion on success stories.

After the pre-project considerations, a working plan was suggested to them and deadlines were fixed. The instructional objectives of working on such projects were explained to the students in detail. They were motivated and monitored even outside the class hours as Littlewood (1984) explains, "motivation is the critical force which determines whether a learner embarks on a task, how much energy he devotes to it and how long he perseveres". The groups were informed to discuss their theme at two stages. The students were guided to develop the content in an organized manner and employ the judicious use of dictions. After training the nuances of delivery, they were equipped with embedding of the right confidence level. All the groups were given 40 minutes to present their project work in front of their classmates as audience. In this duration of 40 minutes, the last 10 minutes were allocated for questions from their peers and feedback from their facilitator. This questioning and feedback session from the peers and facilitator fine-tuned their final stages of the project. After the submission of the project works, the students were enthusiastic to know their assessment and, in some cases, it even led to post discussions.

Brainstorming

A brainstorming session on thinking skill was conducted prior to the demonstration of the case studies. The quote "I think therefore I am" by Rene Descartes, a famous philosopher was discussed to demonstrate the reality of human existence. This fundamental truth was exemplified to the students to explore and give expression to their thought process, as this philosophy equated thinking with being i.e., the very existence and our identity with the active process of thinking domain. Likewise, the exemplification of Jim Davies' fictional cat Garfield claiming, "I eat, therefore I am" was well received by the students and many such humorous quotes were uttered by them and this kind of brainstorming session made the students shed their inhibitions and voice their thoughts, as they could relate to the pun in the quote and the wisdom of thinking and consciousness. Further, the thin demarcation line between creativity and innovation was discussed relating the concept of creativity to inspiration, like the light-bulb moment, the flash of a brilliant idea and innovation to the perspiration involved in taking the idea to the business. This process of commercializing creativity was explicated as the out-of-the box thinking or lateral thinking. In this line, the students' have analysed the case studies of IIT Madhavan, Miticool project, Giant Mirrors of Viganella, Tata Nano, and Mumbai Dabawallas as their project work. Moreover, these case studies are based on a real-life situation in their field of study and it intrigued the students to think laterally and critically.

Results and Discussions

Case study given as classroom activity is like a 'spring board' enabling students to think and respond appropriately. The teaching methodology is geared to this end ensuring 'free play' and 'lively interaction' among students, which in turn, make learning interesting, effective, and useful. However, the enthralling results of the students have validated the point that they have been able to think ahead and take a perspective of the holistic concept of thinking process.

Case Study I

The case study on IIT Madhavan, an engineer from IIT, now a farmer was discussed by the students. Madhavan, an alumnus of IITM had a great passion for agriculture that he resigned his job at ONGC. Madhavan's success story as a farmer explicitly revealed the details of lateral thinking to their classmates. It is notable in this case study that Madhavan was exemplary for his out-of-box thinking as an engineering graduate and for his pursuit of passion relentlessly. He had used the knowledge he obtained in his engineering programme to experiment on innovative ideas and developed creative knowledge in agriculture industry. Though he had a tough time in the initial stages, he was able to learn the experimental techniques in farming, the farming cycle, selling the products, etc. His commitment towards his work made him excel in his endeavours and become a role model to the entrepreneurs. Further, Madhavan's innovative approach of cost effectiveness and mechanization in farming addressed the problems of sparse production and food insecurity in India, and created employment opportunities for the villagers in the agricultural activities. Madhavan's successful venture in his agricultural industry prompted the students to explore the possibilities of pursuing their passionate career by expanding their horizons of thinking abilities.

Case Study II

The second case study on Mansukhbhai Prajapati's Mitticool project, an amazing true story was analysed to portray that a school drop-out achieved a feat that many in the world would envy today. This case study dealt with an innovative solution for a critical problem like electricity and global warming and inspired many youngsters of India. Though there was a lot of opposition to return to his family pottery business, Mansukhbhai came out of his job in a brick factory and started his own business. Mansukhbhai's innovation and risk-taking mentality was reviewed in this particular case study. He tapped the potential of the soil available in his geography and utilized to make the earthenware. He started off with Tavas, clay-mixing machine, and then landed on the idea of converting 'matkas' (pots) to refrigerators. This clay refrigerator comprised an upper part to store water and the bottom unit spaced for fruits, vegetables, and milk. The necessity of the situation made him to innovate this eco-friendly product. Besides, the students acquired the process involved in thinking, initiating from a basic idea or line of thought to bigger productive business outcome. For instance, Mansukhbhai analysed the problem, synthesized the potential information to provide solution, evaluated his ideas with small innovations like Tavas and clay mixing machines, and finally provided an exclusive solution of clay refrigerators. The lateral thinking process was evinced in this case study; for instance, clay was used for making pottery on household items like kitchen utensils, lamps, toys, etc., in India for decades. But Mansukhbhai's lateral thinking idea of using clay for making refrigerators addressed the need of the lower-middle-class community to preserve the products economically.

The former president Dr. A.P.J. Abdul Kalam called him a 'true scientist', as he designed a clay refrigerator that works without electricity and this innovation turned the world's attention to this craftsman in Gujarat. This eco-friendly product won the national award in 2009 from the then president of India. Mansukhbhai's four years of hard labour earned him many accolades and recognition. Mitticool was showcased in a conference organized in the UK in May 2009 and the reputed German firm Bosch and Siemens Hausger te (BSH) showed interest in this product. So, this case study invigorated the students to accomplish their entrepreneurial skill by employing step-by-step thinking process and risk-taking decisions at crucial junctures in their career.

Case Study III

The third case study was on the 'Giant Mirrors of Viganella' that shed sunlight in the winter months to this Italian region of Piedmont that lies at the bottom of a deep valley. This case study provided the insight of thinking process to overcome the real-life situation problems. Due to the mountains, the town Viganella did not get direct sunlight for 83 days a year. To remedy this, in November 2006, the town had set up a giant mirror with adjustable, computer-controlled orientation on the mountainside. For centuries, villagers had accepted their fate, until in 2005 a local engineer and an architect came up with a brilliant idea of using a mirror to reflect sunlight into the village and that was supported by the Mayor who raised 100,000 Euros sharing 540 Euros from each resident. The giant mirror consisting of 40 square metres' mirror, weighting 1.1 tons, was installed on the opposite side of the mountain at an altitude of 1,100 metres. The mirror functions as a heliostat, tracking the sun so that sunlight reflects onto the town square, lighting up an area of 300 square yards for at least six hours a day. After the installation of this giant mirror, a positive change could be observed in the behaviour and mood of the inhabitants. This mirror of Viganella had been attracting the attention of millions of people around the world. Though this lateral idea was initiated by two individuals, it was supported by the residents, and the entire town had worked towards this project. So, this case study was an exhilarating experience for the students setting an example for its innovation and team effort. The successful installation of the mirror encompasses collecting the requisite information like distance to be covered, size of the mirror required, position of the mirrors, its pros and cons, etc. Hence, by working on this case study, the students perceived a holistic approach in the thinking process involved in handling real-life situations and perspiring on the innovative idea from this case study.

Case Studies IV and V

The case studies on Tata Nano and Mumbai Dabbawallas concentrated on addressing the need of a specific group or class of people. The students who worked on the case study of Tata Nano stated that the process of commercializing creativity involved the nuances of getting inspired by out-of-the-box thinking and introducing a car exclusively addressing the need of the middle-class people. They clearly explicated that it does not end up with downsizing another Tata car and a separate team was created for the Nano, to leverage capabilities form the current business and was empowered to discard current notions and innovate a new product forgetting the conventions.

Similarly, the students who worked on the case study of 'Mumbai Dabbawallas' elaborated on the skilful lunchbox delivery system that carried hot lunches from home and restaurants to people at work. The members explicated the lateral thinking concept of tapping the potential of the need of serving people at lunch hours. The group presented their astounding service record of 260,000 transactions in six hours each day, six days a week, and 52 weeks a year (minus holidays). They revealed the capacity of the semiliterate workers who manage themselves in an eco-friendly process without any IT system or cell phones by using a colour-coding system to identify the destination and recipient. They exemplified the legendary service of ordinary workers who were able to achieve extraordinary results and their reliable self-regulatory mechanism in supply chain management occupied a place in the prestigious Harvard Business Review. Further, the group reiterated on the sustaining mission, shared identity, emotional bonding, and synchronizing network that fetched 'Dabbawallas' the recognition of 'Models of Service Excellence'.

These two case studies enthused the students' notion on exploratory business techniques. Tata Nano and Mumbai Dabbawallas targeted the middle-class people and tapped the need of customer convenience and innovated a mechanism to serve their needs. In the 21st Century global workplace, students need to make use of the available opportunities to exhibit their talent to innovate prototypes or products that benefit their target audience. The innovation process could be a simple mechanism or a mega project involving diversified technologies, but they need to understand the concept of commercializing creativity and perspiring to innovate and succeed in their career. These case studies highlighted the lateral thinking process of innovating new product or service, benefitting both the entrepreneur and customer by addressing the needs of the target community.

Similarly, the students comprehensively discussed the case studies like Mumbai Taj Hotel attack, the demolition of Babri Masjid, the violence in Dr. Ambedkar Law College, budget sessions of 2019 and 2020, Sudarsan Patnaik, the sand artist, and Leonardo da Vinci paintings. Most of the student groups opted for lateral and critical thinking case studies, as they were interested to explore on the successful entrepreneurs. The success stories captivated the audience, who were keen and observant in listening to their peers. The questioning and feedback session moulded the students to acquire the nuances of delivery and presentation techniques. The spontaneous question and answer sessions made them fluent in content generation and interact without any constraints and altogether this administering of project-based learning turned out to be a holistic learning experience for the students.

Conclusion

Engineering students learn by doing, as in the case of most of their subjects, they are adopted to experiential mode of learning. Technical English classes also can be theorizing from practical examples say a critical or creative exercise. This chapter has concentrated on providing case study as project work to students and enable their thinking skills. This case study approach is based on the demonstrations, discussions, and presentations of their projects. It has provided the opportunity for the students to think critically and perform on their own, which enhances and hones their skill sets. When language becomes a meaningful instrument in their hands, the students become independent learners with a blend of language competence and performance. Subsequently, this project-based pedagogy will certainly expose the students to the challenges of effectively communicating and efficiently thinking and performing in a given situation. This kind of inquisitive search for learning gives them a working knowledge in multiple domains and trains them to think on one's own feet.

References

Adams, K., Ripper, M., Zander, A., & Mullins, G. (2010). The valuing of creativity in the workplace roles of engineering research graduates. *Paper presented at 3rd international symposium for engineering education, Ireland, 2010.* Ireland: University College Cork.

Basturkmen, H. (2010). *Developing courses in English for specific purposes.* UK: Palgrave Macmillan.

Bloom, B. S., et al. (1956). *Taxonomy of educational objectives handbook I: The cognitive domain.* New York: Longmans.

Butterworth, J., & Thwaites, G. (2013). *Thinking skills: Critical thinking and problem solving.* Second Edition. Cambridge: Cambridge University Press.

Chastain, K. (1976). *Developing second language skills: Theory to practice.* Second Edition. Chicago: Rand McNally.

Cropley, D.H. (2015). Creativity in engineering. In G. E. Corazza & S. Agnoli (Eds.), *Multidisciplinary contributions to the science of creative thinking* (pp. 155–173). London, UK: Springer.

Donnell, J. A., Aller, B.M., Alley, M., & Kedrowicz, A. A. (2011). *AC 2011-1503: Why industry says that engineering graduates have poor communication skills: What the literature says. Paper presented at ASEE annual conference & exposition*, Washington, DC. American Society for Engineering Education.

Eisner, E. (2005). Back to whole. *Educational Leadership, 63*(1), 14–18.

Hall, B. T. (2008, May 25). *Bloom's Revised Taxonomy*. Retrieved from www.esl-school.com/archives/2008/05/blooms_revised_taxonomy.php

Hansraj. (2017). Activity-based teaching – Learning strategy in language. *Scholarly Research Journal for Humanity Science & English Language, 4*(20), 4433–4436.

Ivleva, N. V. (2016). Teaching critical thinking to engineering students through reading profession-oriented texts. *IOP Conference Series: Materials Science and Engineering, 155*, 012022. Doi: 10.1088/1757-899X/155/1/012022

Lavy, I., & Yadin, A. (2013). Soft skills – An importance key for employability in the "Shift to a service driven economy" era. *International Journal of e-Education, e-Business, e-Management and e-Learning, 3*(5), 416–420. Doi: 10.7763/IJEEEE.2013.V3.270

Littlewood, W. (1984). *Foreign and second language learning: Language acquisition research and its implications for the classroom*. Cambridge: Cambridge University Press.

Madhavaiah, U., & Ram, M. V. R. (2016). Enhancing lateral thinking in engineering graduates (Indian Context). *International Journal of Scientific & Engineering Research, 7*(6), 346–350.

Nageswari, R. Ravikumar, B., & Jayamani, T. S. (2016). Building lateral thinking strategies to impart English speaking skills. *Social Sciences & Humanities, 24*(2), 687–700.

Nunan, D. (1988). *The learner-centred curriculum*. Cambridge: University of Cambridge.

Partnership for 21st Century Skills. (2008). *21st century skills, education & competitiveness: A Resource and Policy Guide*. Retrieved from http://www.p21.org

Piirto, J. (2011). *Creativity for 21st century skills: How to embed creativity into the curriculum*. Netherlands: Sense Publishers.

Richards, J. C. (1984). Language curriculum development. *RELC Journal, 15*(1), 1–29.

Sheth, T. D. (2015). Communication skill: A prerequisite for engineers. *International Journal on Studies in English Language and Literature (IJSELL), 3*(7), 51–54.

Simkins, M., Cole, K., Tavalin, F., & Means, B. (2002). *Increasing student learning through multimedia projects*. Alexandria, VA: Association for Supervision and Curriculum Development.

Soule, H. (2020). 21st century skills – An overview. In C. Graham (Eds.), *21st century skills in the ELT classroom – A guide for teachers* (pp. 14–27). UK: Garnet Education.

Tran, M. (2013). Critical thinking for engineers. In ECE Senior Project Design Handbook, *Electrical and computer engineering design handbook: An introduction to electrical and computer engineering and product design*. Massachusetts: Tufts University. Retrieved from https://sites.tufts.edu/eeseniordesignhandbook/2013/an-engineers-path-to-critical-thinking/

Woodrow, L. (2018). *Introducing course design in English for specific purposes*. Oxon: Routledge.

Part V

Approaches and Strategies

Chapter 10

Pedagogical Strategies in Improving ESL Learners' Speaking Proficiency

S. Shantha and R. K. Dharini

Speaking proficiency has become an integral desideratum for engineering students in the present digital and informative era. English Language Proficiency (ELP) facilitates the engineering students to understand the concepts and theories, to write academic reports, to present themselves in interviews, etc. It is a valuable career enhancer and boosts the employment opportunities. Rani and Jayachandran (2014) indicate, "Proficiency in English is a pre-requisite for a successful engineering career and the development of linguistic proficiency in the learner is needed for the spontaneous and appropriate use of language in different situations". The English language is a language of international business, technology, research, and aviation, and the engineers today have to communicate with their counterparts across the globe. Many of the students enter professional courses like engineering without requisite speaking proficiency in English, and the syllabus, teaching, and learning methods do not infuse on improving this crucial need. As school curricula, the curriculum of Technical English I & II in engineering courses concentrates on enriching the students' knowledge about the English language. Though speaking is included in the syllabi of Technical English I & II and Communication Skills Lab, it is totally neglected by the teachers of English. The training and testing of speaking skills are carried out mainly for the examination purpose and not for usage in real life. It is distressing to note that the lack of speaking proficiency of engineering students has immensely affected the general employability of engineering graduates in India.

The recruitment panel of the corporate expects the engineering candidates to be orally proficient in English. As the students are not able to meet this industry requirement, a gap prevails between the corporate expectations and the present academic scenario. This gap has extremely affected the career opportunities of the engineering incumbents in the corporate industry. Nayak (2016) indicates, "there is an urgent need to streamline the English language training in engineering colleges to enhance the employability of the students and make them industry ready". Clement and Murugavel (2015) have explained, "Mere changes in the syllabus will not be able to bring in desired changes unless the English teachers are motivated to enhance their teaching methodologies to bridge the gap between the

DOI: 10.4324/9781003268529-15

college and the workplace". Accordingly, addressing this skills gap, this chapter explicates the influence of pedagogical strategies (PSs) on learning strategies in obliterating the problem of lack of speaking proficiency.

Need for the Study

According to Verma and Bhattacharya (2015),

> A study has found that an overwhelming 97% of engineers in the country cannot speak English, required for high-end jobs in corporate sales and business consulting. Moreover, as per the report based on the study, about 67% of engineers graduating from India's colleges do not possess spoken English skills required for any job in knowledge economy.

In Business Standard, according to the report by Aspiring Minds (2014), a recruiting and HR training firm claims, "25%–35% of engineers are not good at spoken discourses in English, which include day-to-day conversations, official meetings and presentations". The major requisite of engineering students for obtaining job during placements is fluency to speak in English with clarity to convey their thought process. The students who aspire to achieve in their professions require outstanding communication skill, as it equips them to augment their technical and professional skills defining their core competencies to be exhibited (as cited in Panda & Patnayak, 2017).

The workplace demands engineers to have effective speaking skills for giving oral presentations in meetings, conferences, and seminars in front of higher officials or colleagues. Hence, one must know how to express and present one's ideas in order to haul the attention of the audience. Many researchers and reports have revealed that engineering students in India lack spoken proficiency in English, in spite of their long association with English language learning. They lack the ability to generate sentences in English to meet their requirements in real-life situations. Students' inadequacy in spoken English is due to the wide gap between the teaching methodology in the classroom and the use of English in the real life. Mohamed et al. (2020) has stated,

> the previous research has proved the currently inappropriate teaching approaches and the lack of motivation to be the most significant reasons for EFL students' poor speaking proficiency. The lack of motivation is due to the traditional, rigid and tedious teaching techniques adopted in speaking classrooms and partly due to the standardized curriculum of the speaking course where there is no scope for instructors to change their teaching approach or introduce new ideas in the curriculum.

As a result, it is evident that the engineering students are not able to meet the expectations of the corporates when it comes to speaking proficiency.

Research Rationale

The investigation of diverse obstacles faced by the engineering students in the learning process has explicated that oral communicative tasks (OCTs) can be employed to develop the speaking proficiency. Task-based language teaching (TBLT) enables the students to use language in varied contexts in real-life situations. Lochana and Deb (2006) propose, "1. Task-based teaching enhances the language proficiency of the Learners, 2. Tasks encourage learners to participate more in the learning processes". Brown (2008) lists seven principles for designing speaking activities as follows:

- use techniques covering the spectrum of learner needs, i.e., include both accuracy- and fluency-focused activities
- provide intrinsically motivating techniques which appeal to learners' goals and interests
- encourage the use of authentic language in meaningful contexts
- provide appropriate feedback and correction
- capitalize the natural link between speaking and listening
- give learners opportunities to initiate oral communication
- encourage the development of speaking strategies.

(as cited in Jaelani, 2013–2014)

Based on Brown's principles for designing speaking activities, this experimental study has selected 11 motivational strategies of Dornyei to administer the OCTs. Employment of motivational strategies in the implementation of OCTs will reduce the speaking constraints of the learners. The teacher can subdue the psychological barriers by providing motivation, encouragement, confidence, and scope for peer learning in the interactive class room environment. The linguistic problems of the students could be reduced by teacher's feedback and remedial measures. Khan (2010) has investigated the English as a Foreign Language (EFL) learners' strategies and spoken production employing OCTs, with Catalan and Spanish undergraduates as samples, and he has identified the characteristics of tasks to be more influential on the learners' strategy use than their proficiency. The OCTs facilitate the teachers to employ varied motivational strategies as well as facilitate the learners to use appropriate learning strategies. Further, the pedagogical intervention in the implementation of OCTs and language learning strategies (LLS) improves the speaking proficiency of the students.

Speaking Proficiency

Bygate (1987) states,

Speaking means not merely how to assemble sentences in the abstract: we have to produce them and adapt them to the circumstances. This means making decisions rapidly, implementing them smoothly, and adjusting our conversation as unexpected problems appear in our path.

Harmer (2001) indicates that the two important aspects for speech production are knowledge of 'language features' and the ability to process information on the spot, which means 'mental/social processing'. In this era of globalization, it is not enough for engineering students to just have the ability to speak in English. They are expected to be orally proficient in English. Lan (1994) defines speaking proficiency as being able to produce autonomous utterances fluently and accurately that are appropriate to the context of the situation. Bialystok (1978) states that LLS primarily benefit the speaking task. These definitions provide as with the insight on the characteristic aspects of speaking proficiency.

Studies on OCTs and Speaking

Harmer (2001) states three basic reasons for administering speaking tasks to students in the classroom. Firstly, speaking activities provide chances to rehearse the real-life language use. Secondly, speaking tasks provide feedback for both teacher and students. Finally, they provide opportunities to the students to activate the various elements of language they have stored in their brains. A congenial environment needs to be created where the students come forward to perform a task along with their peers. Nashash has investigated the effect of a task-based program for teaching English language productive skills on the development of first-year secondary-grade female students' oral and written skills at a secondary school in Amman (as cited in Yousif, 2017), and the results have revealed that TBLT program has enhanced the learning of communicative speaking and writing skills better than the conventional method of teaching. Aljarf has examined the impact of TBLT on 52 female EFL students and the study has exhibited that the students could speak fluently using correct grammar and pronunciations and could easily generate ideas (as cited in Thanghun, 2012). Zhang (2009) states that students who study EFL usually have limited opportunities to speak English outside the classroom (as cited in Boonkit, 2010). Therefore, the teachers should facilitate adequate situations and tasks for students to develop their speaking skill. Teachers can use various interesting strategies to improve the speaking skills. Students' speaking proficiency can be enhanced by a regular speaking task in the classroom. It requires students and teachers to be engaged in OCTs which will aid them in real-life situations. Mohammadipour and Rashid (2015) have conducted a study to determine the effectiveness of a proposed task-based instruction programme within a cognitive approach in fostering overall speaking proficiency of undergraduate students. Findings of the study have depicted a significant improvement in the overall speaking proficiency of the students who have been exposed to the proposed programme, and have implied the positive impact of the task-based programme in improving the overall speaking proficiency of the undergraduate students. These researches reveal that TBLT offers an opportunity for authentic learning in the classroom and promotes higher level of language proficiency.

Studies on LLS and Speaking

According to Oxford,

> The employment of language learning strategies seems necessary for language learners because speaking the new language often causes the greatest anxiety among other language skills. Oxford explicates "...language learning strategies – specific actions, behaviours, steps, or techniques that students (often intentionally) use to improve their progress in developing L2skills. These strategies can facilitate the internalization, storage, retrieval, or use of the new language. Strategies are tools for the self-directed involvement necessary for developing communicative ability".
> (as cited in Ravi & Sethuraman, 2019)

Ma (2009) has investigated the correlation between language proficiency levels of Chinese college students and their use of strategies, and has found that students with high proficiency level tend to employ more language strategies than those with low proficiency levels. Nguyen (2013) argues that there is a significant relation between the frequency of LLS use and participants' self-rated language abilities for all the four skills. Though many experimental studies have been done for improving the oral proficiency of students through LLS and OCTs separately, PSs that accentuate the impact of LLS and OCTs in improving the speaking proficiency have not been focused. Thus, the present study investigates the role of PSs in administering the OCTs and in the employment of LLS.

Research Questions

1. How the pedagogical strategies influence the employment of language learning strategies?
2. How the use of language learning strategies improves the speaking proficiency of the learners?

Methodology

Research Design

The study will explore the effect of PSs in administering the OCTs and in the use of LLS by the experimental group. Dornyei's (2001) motivational strategies used by the teacher-researcher to administer the OCTs, have facilitated the students to perform the tasks congenially and have enabled them to use learning strategies appropriately. The performance of the students is evaluated in the observation sheets. The pre-task, post-task, and the task performances are assessed using the analytic parameters of spoken language specified in the Common European Framework of Reference (CEFR). The coded data is computed and analysed statistically using SPSS software.

Participants

The participants of the study are first-year Bachelor of Engineering (BE) students of Civil Engineering at the M.A.M College of Engineering and Technology, Tiruchirappalli. The participants (N = 76) are selected based on simple random sampling, in which the samples have been assigned to the control and experimental groups using lottery method (Kothari, 2004). The control and experimental groups consist of 38 students each. Their ages range between 17 and 19 years. The participants comprise 22 females and 54 males. Most of them are from the same background pertaining to their first language, previous educational experience, and learning context. Even though the participants have studied English for around 12 years, they have been identified to be lacking proficiency in the speaking skill in English. In this line, it is assumed that providing appropriate training employing LLS with pedagogical intervention in OCTs will develop the students' speaking proficiency.

Instruments

Questionnaire

A pre-study questionnaire was administered to the participants to know their demographic and academic details. Besides, a post-study questionnaire was used to collect feedback from the participants upon the implementation of the OCTs.

Oral Communicative Tasks (OCTs)

According to Ellis (2003), "A task is a workplan that requires learners to process language pragmatically in order to achieve an outcome that can be evaluated in terms of whether the correct or appropriate propositional content has been conveyed". Self-introduction is used as a pre-task to assess the entry-level speaking proficiency of the control and experimental group. The OCTs have been categorized as initial tasks, core tasks, and supporting tasks. Nine tasks have been devised for the experimental group. The initial tasks administered are 'listing10 activities of a given professional', 'listing of five to do's', 'mentioning associated ideas on a topic', and 'situation-based responses'. The core tasks are 'long answer interview', 'comparing task', 'story completion', 'role-play', and 'group discussion'. The tasks such as 'role-play', 'situation-based responses' resemble real-life situations that are meant to help students confront similar situations in their career. The tasks such as 'story completion' and 'roleplay' are of creative kind and the students could use their imagination and creativity in their performances. The rest of the tasks are informative and the students have to think and present their content. These tasks can be categorized as individual, pair, and group ones. Matera states that pair work and group work provide can be excellent tools to promote student interaction because they are involved in talking to their friends (as cited in Ravi & Sethuraman, 2019). The supporting task on 'short answer sessions' is used to prepare students to respond comprehensibly in English. An impromptu speech on 'the best gift I have ever received' is used as a post-task to validate the exit-level speaking proficiency of the experimental group.

Observation Sheets

The observation method of data collection is distinctive in the sense that it provides the researchers with direct information rather than self-report accounts of the respondents (as cited in Dornyei, 2007). The observations noted in the present study has facilitated the teacher-researcher to identify the constraints faced by the learners, to investigate the factors affecting the speaking proficiency of learners, to analyse the impact of the pedagogical intervention with strategy use on their oral skills and to examine the learning strategies employed in the study. The observation sheets have been used to note down the students' performance in the tasks, say their ability to perform the task, their choice of diction, their sentence construction, coherence in content and correct pronunciation of words. At the end of each task completion, the data from the observation sheets have been transferred to the scoring sheet. The task performance has been scored as per the analytic rubrics specified in the Common European Framework of Reference (CEFR), which is shown in Table 10.1.

Table 10.1 CEFR Speaking Assessment Criteria

Components tested	Weightage of marks
Fluency & coherence	4 marks
Grammatical acceptability	2 marks
Ability to expand the Idea	1 mark
Volume	2 marks
Pronunciation	1 mark
Maximum score	10 marks

Implementation

Dornyei (2001) lists hundreds of strategies in his book on *Motivational Strategies in the Language Classroom,* facilitating the teachers to use them as resources in the language classroom. The teacher-researcher of this study has applied Dornyei's (2001) motivational strategies in administering the OCTs in the classroom and also in making the learners employ learning strategies while performing tasks. The PSs have been customized by the teacher-researcher catering to the needs of the participants of this study as presented in Table 10.2.

The above given strategies have been employed throughout the OCT sessions so as to facilitate the learners to perform the tasks. Wael et al. (2018)

Table 10.2 Pedagogical Strategies Employed in the Study

S. No. of pedagogical strategies	Content of pedagogical strategies
PS-1	Demonstrated and talked about my own enthusiasm for the course material and showed students that I cared about their progress.
PS-2	Showed students that I cared about them and determined to develop their oral proficiency.
PS-3	Promoted the development of group cohesiveness and associated slow learners with their enthusiastic peers (e.g., in group or project work).
PS-4	Raised the learners' intrinsic interest in the L2 learning process and made sure that there were no serious obstacles.
PS-5	Reiterated the role of L2 in the world and encouraged the learners to apply their L2 proficiency in real-life situations.
PS-6	Made learning more enjoyable by breaking the monotony of classroom events and focused on the motivation.
PS-7	Explained the purpose and utility of a task and personalized assignments for everybody.
PS-8	Built learners' confidence by providing regular encouragement and drew their attention to their strengths and abilities.
PS-9	Helped learners accept the fact that they would make mistakes as part of the learning process and taught students learning strategies to facilitate the task performance.
PS-10	Adopted the role of a facilitator throughout the session and provided students with positive information feedback.
PS-11	Monitored student accomplishments and applied continuous assessment that also relies on measurement tools.

state that, "the other factors also influence students' learning strategies in learning speaking such as psychological factor, the role of teacher, task, environment and social factors, the roles of the lecturer as a feedback provider, a participant, and a prompter". The PSs aid the learners to employ various learning strategies during the execution of OCTs. The participants of the study have been enabled to use an array of learning strategies while performing OCTs in the classroom. The teacher researcher has noted the students' use of strategies and the frequency of its usage in the observation sheet. The learning strategies are grouped under different heads for ease of comprehension. The learning strategies have been presented to the participants of the experimental group and been encouraged to employ the strategies in the process of task execution. The students' strategy use in task performance have been duly noted by the teacher researcher in the observation sheet. The observations by the researcher have been converted in to a 5-point Likert scale values (5-always, 4-often, 3-sometimes, 2-rarely and 1-never). As the role of a teacher is instrumental in facilitating the students in OCT classes to attain L2 oral proficiency, the researcher has employed the strategies as mentioned in Dornyei (2001) in executing the tasks. At the outset, the teacher demonstrated the need for the implementation of OCTs to the experimental group in improving their spoken proficiency. The learners have also been informed of how she cared about their progress and has been determined to develop their oral proficiency, which is a pre-requisite for their career as prospective engineers. The participants have also been briefed on a general idea about the OCT program and its significance before it was implemented to them.

The teacher-researcher has promoted group cohesiveness by grouping the slow learners with their enthusiastic peers in pair and group tasks in order to instigate confidence in themselves. The facilitator has sustained the interest of the participants in the L2 learning process and has made sure that they did not encounter any inhibition in approaching her for clarifications during their learning process throughout the course. The teacher has also emphasized the use of speaking in English for their career prospects and has made them practise speaking in their real-life situation so as to develop their oral proficiency. The facilitator has ensured a conducive environment in the classroom and has motivated the participants to tone down the psychological constraints faced by the participants. The facilitator has provided encouragement to the participants and has made them aware of their strengths and abilities, in order to build their confidence. She has expressed that mistakes are a part of the learning process and has taught them to use learning strategies appropriately in performing the tasks.

The classroom has been learner-centred, where the focus has been on the active involvement and interaction of the learners during the learning process. The participants have felt free in taking up the tasks and gain interest and confidence to perform the task. The facilitator has made sure to explain the purpose and the utility of a task, before executing it in the classroom.

She has organized demonstrative sessions to enhance the understanding of tasks and subdued their fears and inhibitions. Through interactive classes and feedback sessions, the teacher has subdued the linguistic constraints of the students. The teacher has guaranteed that all the students would begiven equal opportunities. Every learner is unique in his/her performance and so the participants have been given personalized assignments. Dornyei (2001) states that the teacher is required to acknowledge the learners' different needs and learning style, understand their needs and goals, communicate trust and respect for them, and give feedback on their learning. All these would aid in improving their self-esteem and confidence (as cited in Ravi & Sethuraman, 2019).

Data Analysis

The data was collected and analysed using SPSS software. One-way ANOVA was computed to find out whether there is any significant difference between the use of learning strategies of the experimental group with respect to the implementation of PSs. Further, correlation analysis was computed in order to identify the relation between the use of learning strategy and the speaking proficiency of the participants of the study.

Results

It is obvious that all the PSs employed by the facilitator had an impact on the cognitive aspect of learning strategies. According to Lan and Lucas (2015), "the more positive attitude and the more positive motivation students have the better language learning strategies they use in their own language learning".

As a part of PS-I, the facilitator has demonstrated and talked about her own enthusiasm for designing the course material and explicated to the students that she cared about their progress. The group mean of the variables is calculated and 'F' value is determined to find out whether there is notable difference among the learning strategies employed by the experimental group with respect to the implementation of PS-I. It is comprehended from the above table that F is significant at 0.01 level showing a significant difference among the use of learning strategies namely cognitive, affective, and resource-based strategies with respect to the employment of PS-I. Hence, it is inferred that the pedagogical implication of the OCTs administered by the facilitator created an impact on the cognitive, affective, and resource-based strategies. Mistar et al. (2014) have examined the strategies employed by EFL learners for learning speaking skill and found that "resources-based strategies were found to be the most intensively deployed strategies". In congruence with this, the results in Table 10.3 indicate that cognitive and affective strategies play a major role in improving the speaking proficiency of the learners.

Table 10.3 One-Way ANOVA for Pedagogical Strategy-I

Learning strategies		Sum of squares	DF	Mean square	F	Sig.
Self-improvement	Between Groups	394.491	2	197.245	2.802	0.074 (NS)
	Within Groups	2463.404	35	70.383		
	Total	2857.895	37			
Compensation	Between Groups	70.504	2	35.252	2.117	0.136 (NS)
	Within Groups	582.760	35	16.650		
	Total	653.263	37			
Cognitive	Between Groups	4193.970	2	2096.985	38.713	0.01
	Within Groups	1895.846	35	54.167		
	Total	6089.816	37			
Interpersonal	Between Groups	4.935	2	2.468	0.116	0.891 (NS)
	Within Groups	746.538	35	21.330		
	Total	751.474	37			
Affective	Between Groups	1961.435	2	980.718	19.727	0.01
	Within Groups	1740.038	35	49.715		
	Total	3701.474	37			
Resource-based	Between Groups	1304.881	2	652.440	24.768	0.01
	Within Groups	921.962	35	26.342		
	Total	2226.842	37			

As per the plan of PS-II, the facilitator has explained to the learners that she has been determined to improve their speaking proficiency in PS-II. The group mean of the variables is calculated and 'F' value is determined to find out whether there is noticeable difference among the learning strategies employed by the experimental group with reference to the implementation of PS-II. Table 10.4 reveals that there is a significant difference among the use of learning strategies namely compensatory, cognitive, affective, and resource-based strategies ($p < 0.01$) with reference to the application of PS-II. According to Martina et al. (2020), "competent speakers used much more compensation strategies than poor speakers". It suggests that compensation strategies are extremely useful as guidance to avoid communication gap in speaking activities. The learners' use of strategies reveals the fact that the facilitator has cared in the progress of their learners. It is inferred that the pedagogical intervention impacts the learners' employment of the learning strategies.

Table 10.4 One-Way ANOVA for Pedagogical Strategy-II

Learning strategies		Sum of squares	DF	Mean square	F	Sig.
Self-improvement	Between Groups	136.010	2	68.005	0.874	0.426 (NS)
	Within Groups	2721.885	35	77.768		
	Total	2857.895	37			
Compensation	Between Groups	107.930	2	53.965	3.464	0.05
	Within Groups	545.333	35	15.581		
	Total	653.263	37			
Cognitive	Between Groups	2550.542	2	1275.271	12.611	0.01
	Within Groups	3539.274	35	101.122		
	Total	6089.816	37			
Interpersonal	Between Groups	43.044	2	21.522	1.063	0.356 (NS)
	Within Groups	708.430	35	20.241		
	Total	751.474	37			
Affective	Between Groups	1191.311	2	595.655	8.305	0.01
	Within Groups	2510.163	35	71.719		
	Total	3701.474	37			
Resource-based	Between Groups	868.042	2	434.021	11.180	0.01
	Within Groups	1358.800	35	38.823		
	Total	2226.842	37			

As per the PS-III, the facilitator has promoted the development of group cohesiveness and grouped slow learners with their enthusiastic peers. The group mean of the variables is calculated and 'F' value is determined to find out whether there is remarkable difference among the learning strategies employed by the experimental group with respect to the implementation of PS-III. Table 10.5 reveals that F is significant at 0.01 level showing a significant difference among the learning strategies, i.e., self-improvement, compensatory, cognitive, affective, and resource-based strategies pertaining to the employment of PS-III. *Mistar and Umamah (2014) have stated,* "four strategy types – interactional-maintenance, self-improvement, compensation, and memory strategies – greatly contribute to the speaking proficiency". It is obvious that the facilitator's effort in building the group cohesiveness has prompted the learners to use the learning strategies effectively.

Table 10.5 One-Way ANOVA for Pedagogical Strategy-III

Learning strategies		Sum of squares	DF	Mean square	F	Sig.
Self-improvement	Between Groups	866.380	2	433.190	7.613	0.01
	Within Groups	1991.515	35	56.900		
	Total	2857.895	37			
Compensation	Between Groups	111.021	2	55.510	3.583	0.05
	Within Groups	542.242	35	15.493		
	Total	653.263	37			
Cognitive	Between Groups	4089.554	2	2044.777	35.779	0.01
	Within Groups	2000.261	35	57.150		
	Total	6089.816	37			
Interpersonal	Between Groups	106.504	2	53.252	2.890	0.069 (NS)
	Within Groups	644.970	35	18.428		
	Total	751.474	37			
Affective	Between Groups	2403.474	2	1201.737	32.404	0.01
	Within Groups	1298.000	35	37.086		
	Total	3701.474	37			
Resource-based	Between Groups	849.433	2	424.717	10.792	0.01
	Within Groups	1377.409	35	39.355		
	Total	2226.842	37			

As per PS-IV, the facilitator has raised the learners' intrinsic interest in the L2 learning process and has made sure that there were no serious obstacles. According to Wilona et al. (2010), "Having Intrinsic Motivation in the process of learning a foreign/second language helps people to achieve better speaking proficiency". The group mean of the variables is calculated and 'F' value is determined to find out whether there is appreciable difference among the learning strategies employed by the experimental group in connection with the implementation of PS-IV. It is understood from Table 10.6 that there is a significant variance among the use of learning strategies such as cognitive, affective, and resource-based strategies (p < 0.01) with respect to the employment of PS-IV. Rahman and Maarof (2016) has proposed, "motivation correlated positively with all types of language learning strategies". It is inferred that the learners have faced some initial obstacles despite

Table 10.6 One-Way ANOVA for Pedagogical Strategy-IV

Learning strategies		Sum of squares	DF	Mean square	F	Sig.
Self-improvement	Between Groups	59.224	2	29.612	0.370	0.693 (NS)
	Within Groups	2798.671	35	79.962		
	Total	2857.895	37			
Compensation	Between Groups	36.964	2	18.482	1.050	0.361 (NS)
	Within Groups	616.299	35	17.609		
	Total	653.263	37			
Cognitive	Between Groups	4324.875	2	2162.438	42.883	0.01
	Within Groups	1764.941	35	50.427		
	Total	6089.816	37			
Interpersonal	Between Groups	29.015	2	14.508	0.703	0.502 (NS)
	Within Groups	722.458	35	20.642		
	Total	751.474	37			
Affective	Between Groups	2494.565	2	1247.282	36.171	0.01
	Within Groups	1206.909	35	34.483		
	Total	3701.474	37			
Resource-based	Between Groups	1001.407	2	500.704	14.301	0.01
	Within Groups	1225.435	35	35.012		
	Total	2226.842	37			

the teacher's encouragement. But subsequently the learners have been able to overcome the impediments by their active participation in the tasks.

As per the PS-V, the facilitator has reiterated the role of L2 in the world and encouraged the learners to apply their L2 proficiency in real-life situations. The group mean of the variables is calculated and 'F' value is determined to find out whether there is prominent difference among the learning strategies employed by the experimental group in relation to the implementation of PS-V. Table 10.7 shows that there is a significant difference among the use of learning strategies, i.e., self-improvement, compensatory, cognitive, affective, and resource-based strategies (p < 0.01) with regard to the application of PS-V. It is inferred that the learners have comprehended the significance of L2 in real-life situations, as is evident from the application of substantial number of learning strategies.

Table 10.7 One-Way ANOVA for Pedagogical Strategy-V

Learning strategies		Sum of squares	DF	Mean square	F	Sig.
Self-improvement	Between Groups	1336.481	2	445.494	9.956	0.01
	Within Groups	1521.414	35	44.747		
	Total	2857.895	37			
Compensation	Between Groups	133.050	2	44.350	2.899	0.05
	Within Groups	520.213	35	15.300		
	Total	653.263	37			
Cognitive	Between Groups	4341.011	2	1447.004	28.132	0.01
	Within Groups	1748.804	35	51.435		
	Total	6089.816	37			
Interpersonal	Between Groups	136.861	2	45.620	2.524	0.074 (NS)
	Within Groups	614.613	35	18.077		
	Total	751.474	37			
Affective	Between Groups	2510.774	2	836.925	23.898	0.01
	Within Groups	1190.700	35	35.021		
	Total	3701.474	37			
Resource-based	Between Groups	1063.607	2	354.536	10.363	0.01
	Within Groups	1163.235	35	34.213		
	Total	2226.842	37			

The facilitator has focused on the motivational aspect as a part of PS-VI and has made learning more interesting by breaking the monotony of classroom events. The group mean of the variables is calculated and 'F' value is determined to find out whether there is appreciable difference among the learning strategies employed by the experimental group with regard to the implementation of PS-VI. Table 10.8 indicates that there is a significant difference among the use of learning strategies such as self-improvement, compensatory, cognitive, affective, and resource-based strategies ($p < 0.01$) with reference to the employment of PS-VI. Sien et al. (2018) have argued, "types of speaking task and instrumental motivation had great impact on the selections of language learning strategies among learners". It is explicit that the facilitator's motivation has had a positive impact on the learners to make considerable use of most of the learning strategies not only in their task performances but also in their real-life situations.

Table 10.8 One-Way ANOVA for Pedagogical Strategy-VI

Learning Strategies		Sum of squares	DF	Mean square	F	Sig.
Self-improvement	Between Groups	1243.209	2	621.605	13.474	0.01
	Within Groups	1614.686	35	46.134		
	Total	2857.895	37			
Compensation	Between Groups	151.263	2	75.632	5.273	0.01
	Within Groups	502.000	35	14.343		
	Total	653.263	37			
Cognitive	Between Groups	3128.406	2	1564.203	18.487	0.01
	Within Groups	2961.410	35	84.612		
	Total	6089.816	37			
Interpersonal	Between Groups	69.397	2	34.699	1.781	0.183 (NS)
	Within Groups	682.076	35	19.488		
	Total	751.474	37			
Affective	Between Groups	1851.188	2	925.594	17.509	0.01
	Within Groups	1850.286	35	52.865		
	Total	3701.474	37			
Resource-based	Between Groups	880.823	2	440.412	11.452	0.01
	Within Groups	1346.019	35	38.458		
	Total	2226.842	37			

The facilitator has explained the purpose and utility of tasks as a part of PS-VII and assigned personalized tasks for every learner. The group mean of the variables is calculated and 'F' value is determined to find out whether there is obvious difference among the learning strategies employed by the experimental group with regard to the implementation of PS-VII. Table 10.9 displays that F is significant at 0.01 levels showing a significant difference among the use of all learning strategies except interpersonal strategy with respect to the employment of PS-VII. It is inferred that the personalized assignments administered by the facilitator enabled the learners to employ almost all the learning strategies effectively. Wael et al. (2018) have propounded that the students from English department used memory, metacognitive, affective, compensation, and cognitive strategies in speaking performance. The personalized assignments have fostered their language potential and have increased their efficiency in their individual performances.

Table 10.9 One-Way ANOVA for Pedagogical Strategy-VII

Learning Strategies		Sum of squares	DF	Mean square	F	Sig.
Self-improvement	Between Groups	1227.605	2	409.202	8.534	0.01
	Within Groups	1630.290	35	47.950		
	Total	2857.895	37			
Compensation	Between Groups	167.716	2	55.905	3.915	0.01
	Within Groups	485.547	35	14.281		
	Total	653.263	37			
Cognitive	Between Groups	3929.789	2	1309.930	20.619	0.01
	Within Groups	2160.026	35	63.530		
	Total	6089.816	37			
Interpersonal	Between Groups	131.330	2	43.777	2.400	0.085 (NS)
	Within Groups	620.143	35	18.240		
	Total	751.474	37			
Affective	Between Groups	2294.137	2	764.712	18.475	0.01
	Within Groups	1407.336	35	41.392		
	Total	3701.474	37			
Resource-based	Between Groups	1163.006	2	387.669	12.390	0.01
	Within Groups	1063.836	35	31.289		
	Total	2226.842	37			

The facilitator has improved the confidence level of the learners as a part of PS-VIII by providing regular encouragement and focused their attention to their strengths and abilities. The group mean of the variables is calculated and 'F' value is determined to find out whether there is appreciable difference among the learning strategies employed by the experimental group with reference to the implementation of PS-VIII. Table 10.10 shows that there is a significant difference among the use of learning strategies namely self-improvement, compensatory, cognitive, interpersonal, affective, and resource-based strategies ($p < 0.01$) with regard to the application of PS-VIII. The findings of the study conducted by Safari & Fitriati (2016) have revealed,

(1) Learners with high speaking performance used all kinds of strategies in learning speaking. They employed those strategies in the equal degree of frequency. (2) Learners with low speaking performance

Table 10.10 One-Way ANOVA for Pedagogical Strategy-VIII

Learning Strategies		Sum of squares	DF	Mean square	F	Sig.
Self-improvement	Between Groups	1225.006	2	408.335	8.502	0.01
	Within Groups	1632.889	35	48.026		
	Total	2857.895	37			
Compensation	Between Groups	186.096	2	62.032	4.515	0.01
	Within Groups	467.167	35	13.740		
	Total	653.263	37			
Cognitive	Between Groups	3933.316	2	1311.105	20.671	0.01
	Within Groups	2156.500	35	63.426		
	Total	6089.816	37			
Interpersonal	Between Groups	170.807	2	56.936	3.334	0.05
	Within Groups	580.667	35	17.078		
	Total	751.474	37			
Affective	Between Groups	2334.140	2	778.047	19.347	0.01
	Within Groups	1367.333	35	40.216		
	Total	3701.474	37			
Resource-based	Between Groups	1169.287	2	389.762	12.531	0.01
	Within Groups	1057.556	35	31.105		
	Total	2226.842	37			

usually used cognitive, metacognitive and social strategies. They also did not apply those strategies in equal degree of frequency. (3) Learners with high speaking performance used strategies more dominantly and actively than those with low speaking performers. (4) Learners with high speaking performance seemed to have higher motivation than low speaking performance. This case influences the application of those strategies. (5) Problems encountered by learners with low speaking performance were in the application of strategies.

It is evident that the encouragement provided by the teacher motivated the learners to use the learning strategies and perform the tasks successfully.

As per PS-IX, the facilitator has helped learners to accept the fact that they would make mistakes as a part of their learning process and has taught students learning strategies to facilitate the task performance. The group

Table 10.11 One-Way ANOVA for Pedagogical Strategy-IX

Learning strategies		Sum of squares	DF	Mean square	F	Sig.
Self-improvement	Between Groups	1236.666	2	412.222	8.645	0.01
	Within Groups	1621.229	35	47.683		
	Total	2857.895	37			
Compensation	Between Groups	138.377	2	46.126	3.046	0.05
	Within Groups	514.887	35	15.144		
	Total	653.263	37			
Cognitive	Between Groups	3722.986	2	1240.995	17.827	0.01
	Within Groups	2366.829	35	69.613		
	Total	6089.816	37			
Interpersonal	Between Groups	149.690	2	49.897	2.819	0.054 (NS)
	Within Groups	601.784	35	17.700		
	Total	751.474	37			
Affective	Between Groups	2229.245	2	743.082	17.161	0.01
	Within Groups	1472.229	35	43.301		
	Total	3701.474	37			
Resource-based	Between Groups	950.725	2	316.908	8.443	0.01
	Within Groups	1276.117	35	37.533		
	Total	2226.842	37			

mean of the variables is calculated and 'F' value is determined to find out whether there is remarkable difference among the learning strategies employed by the experimental group in connection with the implementation of PS-IX. Table 10.11 reveals that the F is significant at 0.01 level showing a significant difference among the learning strategies, i.e., self-improvement, compensatory, cognitive, affective, and resource-based strategies with respect to the employment of PS-IX. It is evident that the learners have realized that errors are part of their learning process, which is a component of self-improvement and affective strategies. The learners have improved in the application of the learning strategies at appropriate junctures and their speaking proficiency gradually progressed in the due course.

The facilitator has provided encouragement and positive feedback as a part of PS-IX that created a good impact among the learners. The group mean of the variables is calculated and 'F' value is determined to find out

Table 10.12 One-Way ANOVA for Pedagogical Strategy-X

Learning strategies		Sum of squares	DF	Mean square	F	Sig.
Self-improvement	Between Groups	57.205	2	57.205	0.735	0.397 (NS)
	Within Groups	2800.690	35	77.797		
	Total	2857.895	37			
Compensation	Between Groups	50.046	2	50.046	2.987	0.093 (NS)
	Within Groups	603.217	35	16.756		
	Total	653.263	37			
Cognitive	Between Groups	3662.169	2	3662.169	54.307	0.01
	Within Groups	2427.646	35	67.435		
	Total	6089.816	37			
Interpersonal	Between Groups	15.584	2	15.584	0.762	0.388 (NS)
	Within Groups	735.890	35	20.441		
	Total	751.474	37			
Affective	Between Groups	2162.088	2	2162.088	50.562	0.01
	Within Groups	1539.386	35	42.761		
	Total	3701.474	37			
Resource-based	Between Groups	713.300	2	713.300	16.966	0.01
	Within Groups	1513.542	35	42.043		
	Total	2226.842	37			

whether there is notable difference among the learning strategies employed by the experimental group with respect to the implementation of PS-X. Table 10.12 shows that there is a significant variance among the use of learning strategies namely cognitive, affective, and resource-based strategies ($p < 0.01$) pertaining to the application of PS-X. This congenial classroom environment has made the students approach the facilitator for clarifications. It has been explicitly indicated to students that the administered OCTs to them were observed for their progress in proficiency attainment and not considered as an assessment criterion. This positive gesture has motivated the learners in carrying out the tasks successfully. In addition, the teacher's positive feedback has assisted their progress in task performance.

The facilitator has monitored learners' accomplishments and applied continuous assessment that relies on measurement tools as per PS-XI. The group mean of the variables is calculated and 'F' value is determined to

Table 10.13 One-Way ANOVA for Pedagogical Strategy-XI

Learning strategies		Sum of squares	DF	Mean square	F	Sig.
Self-improvement	Between Groups	763.933	2	381.967	6.384	0.01
	Within Groups	2093.961	35	59.827		
	Total	2857.895	37			
Compensation	Between Groups	55.628	2	27.814	1.629	0.211 (NS)
	Within Groups	597.635	35	17.075		
	Total	653.263	37			
Cognitive	Between Groups	4248.956	2	2124.478	40.392	0.01
	Within Groups	1840.860	35	52.596		
	Total	6089.816	37			
Interpersonal	Between Groups	62.672	2	31.336	1.592	0.218 (NS)
	Within Groups	688.802	35	19.680		
	Total	751.474	37			
Affective	Between Groups	2483.918	2	1241.959	35.702	0.01
	Within Groups	1217.556	35	34.787		
	Total	3701.474	37			
Resource-based	Between Groups	1200.852	2	600.426	20.483	0.01
	Within Groups	1025.990	35	29.314		
	Total	2226.842	37			

find out whether there is prominent difference among the learning strategies employed by the experimental group in relation to the implementation of PS-XI. Table 10.13 indicates that there is a significant difference among the use of learning strategies namely self-improvement, cognitive, affective, and resource-based strategies ($p < 0.01$) with reference to the employment of PS-XI. It is inferred that the teacher's continuous monitoring and assessment have aided her to provide feedback for the learners' improvement in their oral performance. The PSs have influenced the learning strategies substantially that the learners have been aware of the significance of applying the learning strategies at appropriate junctures on viewing the initiatives taken by their facilitator (as a part of PSs) say maintaining congenial class atmosphere, forming group cohesiveness, enabling them to realize that errors were actually part of learning process, etc. The influence of PSs over the learning strategies explicates the need for improving the use of learning strategies with the intervention of teacher. The extensive use of learning strategies confirms the remarks made by Bygate (1987), "the use of formulaic expressions, hesitation devices, self-correction, rephrasing and repetition can help learners become more fluent and, sound natural".

Correlation Analysis between Speaking Proficiency and Learning Strategies

It is evident from Table 10.14 that the use of interpersonal strategies correlates with speaking components such as grammar (0.013), content (0.000), volume (0.046), and pronunciation (0.021). The learners always preferred to work with others in groups and pairs; they sought help from their peers without any inhibition in group tasks. The enthusiastic learners came forward to help their friends, when they were stuck in the middle of their task performance. The self-improvement and compensatory strategies correlate with the grammar component (0.003 and 0.006). The compensatory strategies, say using contextual clues and gestures, making intelligent guesses, enabled the participants to accomplish the tasks given to them. The

Table 10.14 Correlation Analysis of Speaking and Learning Strategy Use

	Flu.	Gra.	Cont.	Vol.	Pron.	SIS	ComS	CogS	IS	AS	RS
Flu.											
Gra.	0.46**										
	0.00										
Cont.	0.25	0.42**									
	0.13	0.01									
Vol.	0.31	0.32	0.41*								
	0.06	0.05	0.01								
Pron	0.45**	0.43**	0.63**	0.31							
	0.01	0.01	0.00	0.06							
SIS	0.32	0.47**	0.20	0.29	0.14						
	0.05	0.00	0.22	0.08	0.40						
CmS	0.29	0.44**	0.19	0.25	0.11	0.54**					
	0.08	0.01	0.26	0.13	0.52	0.00					
CgS	−0.09	0.11	0.06	0.31	−0.25	0.16	0.40*				
	0.58	0.50	0.73	0.06	0.13	0.33	0.01				
IS	0.15	0.40**	0.63**	0.33*	0.37*	0.37*	0.41*	0.40*	0.18		
	0.39	0.01	0.00	0.05	0.02	0.02	0.01	0.01	0.28		
AS	−0.15	0.09	0.01	0.19	−0.29	−0.29	0.24	0.46**	0.87**	0.23	
	0.38	0.59	0.98	0.25	0.08	0.08	0.15	0.00	0.00	0.16	
RS	0.02	0.32	0.20	0.33*	−0.14	0.36*	0.51**	0.85**	0.40*	0.82**	
	0.91	0.05	0.22	0.04	0.40	0.03	0.00	0.00	0.01	0.00	

** Correlation is significant at the 0.01 level (two-tailed)
* Correlation is significant at the 0.05 level (two-tailed)
Note: Flu. – Fluency and coherence, Gra. – Grammar, Cont. – Content, Vol – Volume, Pron – Pronunciation, SIS – Self-improvement Strategy, Com S – Compensatory Strategy, Cog.S – Cognitive Strategy, IS – Interpersonal Strategy, AS – Affective Strategy, RS – Resource-based Strategy.

self-improvement strategies such as looking for ways to improve and not to repeat mistakes, learning to correct from errors, enabled the learners to be conscious of making error-free performances in the tasks. Further, the significance of learning grammar in L2 environment is pertinent in motivating the students to acquire grammatically acceptable utterances. The teacher's observations and the positive tendency of the learners have created an impact in effective execution of the OCTs and enhanced their speaking skill.

The use of resource-based strategies correlates with volume (0.042). This is an indication that the learners gained confidence, as is evident in their raised volume of utterances in their task performances. Once the learners have started to use resources like dictionary and their peer models, they gained boldness and spirit that resulted in raising their volume in speech. The interpersonal strategies, i.e., the influence of teachers and peers have played a vital role in aiding the learners when they searched words in their presentation, required correct usage of phrases and sentences. The participants have exhibited their interpersonal skills such as seeking help from better learners and coming forward in helping their peers when they stuck in the middle of their performances. The employment of affective and cognitive strategies does not correlate with speaking components. The p values of affective strategy and cognitive strategy with respect to the learners' speaking skills are higher than the significant value 0.05. The affective strategies have assisted few enthusiastic learners to deal with their emotions and attitudes. Participant 23 has been totally unaffected, while her peers mocked at her performance. She has been able to sustain the confidence level throughout the course and continued to make bold attempts in performing the task. But majority of the learners have been filled with anxiety when they made the presentation. The learners have performed the speaking tasks for the first time and they did not intend and have not accustomed to use the cognitive strategies. It is a welcoming initiative for a learner to analyse one's own speech, monitor one's speech, or take charge of one's learning.

Moreover, when the learners are habituated to customize and use such strategies, it would create positive impact. *Gani et al. (2015) have propounded,*

> high performance speaking students had better balance in using all kinds of learning strategies (memory, cognitive, compensatory, metacognitive, affective, and social) for enhancing their speaking skills; the same could not be found with low performance speaking students. Besides, the high-performance students employed more learning strategies consciously and appropriately compared to the low performance students.

Certain strategies have correlated positively with speaking components to enhance their proficiency.

Discussion

Research Question 1

The results of one-way ANOVA for each set of PS suggest that the PSs greatly influence the choice of learning strategies. The PSs have enabled the learners to use appropriate learning strategies improving the speaking proficiency of the learners. The objective of pedagogical intervention in this study has been to develop the speaking ability of the learners and to facilitate the students to achieve language proficiency through interactive participation and application of learning strategy. The learners are enabled to choose the strategies themselves, rather than teacher prescribing the strategies. There has been a significant improvement in speaking components in the post-task. It may be because of their effectiveness in undertaking the tasks despite their use of learning strategies. The pertinent requisite is the involvement of the participants in applying learning strategies, and it cannot be coerced on the learners. The teacher researcher has presented the learning strategies with their respective rationale and enabled the learners to use them independently. The PSs have influenced the learning strategies substantially that the learners have been aware of the significance of applying the learning strategies at appropriate junctures on viewing the initiatives taken by their facilitator (as a part of PSs), say maintaining congenial class atmosphere, forming group cohesiveness, enabling them to realize that errors have been actually part of learning process, etc. The influence of PSs over the learning strategies explicates the need for improving the use of learning strategies with the intervention of teacher. According to Kustati (2012), "the use of speaking-related LLS and learning motivation are demonstrably related to the students' achievement and proficiency". The learners' use of strategies in their task execution have helped them to progress in their task performance and overcome their constraints in speaking. The teacher's use of each PS encouraged the experimental group to apply a wide range of learning strategies in their oral practice in the classroom interaction. The students have been made to interact with peers to overcome their constraints in speaking performance. Their participation has increased their confidence level to make oral presentation.

Research Question 2

The teacher has enabled the learners to use learning strategies to overcome the constraints faced during their task performances. Salikin (2017) states that LLS play a significant part in the learning process and achievement of high achiever students by influencing their way of thinking and motivation. Learning strategies are employed to reduce the obstacles in the learning process. Learning strategies improve the confidence of high achiever students. The experimental group of the study attempted to use the learning strategies in their task performances. It has been noted that most of the participants applied interpersonal strategies frequently, while performing the OCTs. Correlation analysis between the use of learning strategies and speaking proficiency shows that the interpersonal strategies correlate well with the speaking components. Utama and Shabir (2017) have argued, "high achiever students dominantly used compensation and social strategies for learning speaking". The strategies such as seeking help from learners and facilitator, approaching teacher for translating words from mother tongue and enjoying working in group, have aided the participants to perform in their respective tasks. It has been observed that the enthusiastic learners have tried to apply certain self-improvement strategies, say they started to learn from their correction of errors, they became conscious not to repeat their errors, and also, they tried to learn from fluent speakers in the class. These learners have handled affective strategies well, as they did not seem to consider any mocking comments of their peers and they also were not disturbed by their errors. The rest of the participants seemed to be reluctant in making bold attempts in their performance and they often sought the facilitator's help and encouragement while performing the OCTs. The participants have also applied compensatory strategies during their tasks by making guesses from contextual clues, replacing words with gestures, etc. The resource-based strategies such as using dictionary in the class and generating sentences from peers' speech enabled highly motivated students to try new words and perform their tasks effectively. The high achievers have resorted to cognitive strategies such as monitoring and analysing her/his own speech, learning to vary her/his language according to situation, finding their own way of taking charge of language. Liansari (2016) states, "successful English learners used both direct and indirect strategies in learning to speak English".

Conclusion

Students' English language is considered successful, if they are proficient to speak and communicate their thoughts effectively in English language. However, it has been pointed out by various agencies how lack of spoken English skills affects the future career prospects of engineering students. Besides, the goal of teaching speaking skills is to attain oral communicative proficiency. Thus, to improve the oral proficiency of the students, this experimental study has employed motivational strategies of Dornyei to make students use appropriate learning strategies while performing the OCTs. The PSs have been employed to impact the cognitive aspects of learning strategies and thereby improve the efficacy of OCTs in effectuating the students' speaking proficiency. The results of the study emphasize that the PSs have influenced the learning strategies substantially that the learners have been enabled to apply learning strategies at appropriate junctures. The results also reveal that the learners have accomplished in improving their oral proficiency and they have been empowered to apply learning strategies in their spoken performance. The findings of the study recommend the use of motivational strategies as pedagogical intervention to overcome the major impediments faced by the engineering students in the ESL context. This study makes it evident that the implementation of OCTs aided with effective tools like PSs and learner strategies have made the experimental group students overcome their speaking constraints and outperform the control group students. This study has addressed one of the long ongoing issues of improving the speaking proficiency of the engineering graduates in this globalized era. This experimental study explicitly indicates that the speaking proficiency of the students can be improved by devising OCTs, and it also calls upon English teachers' attention towards their pertinent role of improving language proficiency of their learners in ESL context.

References

Aspiring Minds. (2014). *National employability report—Engineers* [annual report]. Retrieved from http://www.aspiringminds.com/sites/default/files/National%20 Employability%20Report%20%20Engineers%2C%20Annual%20Report%20 2014.pdf (pp. 1–63).

Bialystok, E. (1978). A theoretical model of second language learning1. *Language Learning, 28*(1), 69–83. doi: 10.1111/j.1467-1770.1978.tb00305.x

Boonkit, K. (2010). Enhancing the development of speaking skills for non-native speakers of English. *Procedia – Social and Behavioral Sciences, 2*(2), 1305–1309. doi: 10.1016/j.sbspro.2010.03.191

Brown, H. D. (2008). *Teaching by principles. An interactive approach to language pedagogy. Business standard* (3rd ed.) White Plains, NY: Pearson Longman. Retrieved from http://www.businessstandard.com.

Bygate, M. (1987). *Speaking.* Oxford: Oxford University Press.

Clement, A., & Murugavel, T. (2015). English for employability: A case study of the English language training needs analysis for engineering students in India. *English Language Teaching, 8*(2), 116–125. doi: 10.5539/elt.v8n2p116

Dornyei, Z. (2001). *Motivation strategies in the language classroom.* Cambridge: Cambridge University Press.

Dornyei, Z. (2007). *Research methods in applied linguistics.* Cambridge: Cambridge University Press.

Ellis, R. (2003). *Task-based language learning and teaching.* Oxford: Oxford University Press.

Gani, S. A., Fajrina, D., & Hanifa, R. (2015). Students' learning strategies for developing speaking ability. *Studies in English Language and Education, 2*(1), 16–28. doi: 10.24815/siele.v2i1.2232

Harmer, J. (2001). *The practice of English language teaching* (3rd ed.) Harlow, England: Pearson Education.

Jaelani, Z. (2013–2014). Efforts to improve students' speaking skills through communicative activities: A classroom action research at Grade VIII of SMP. *Yogyakarta* [Bachelor's Thesis]. Retrieved from https://core.ac.uk/download/ pdf/33516315.pdf, *N8.*

Kassem, M. A. M. (2018). Improving EFL students' speaking proficiency and motivation: A hybrid problem-based learning approach. *Theory and Practice in Language Studies, 8*(7), 848–859. doi: 10.17507/tpls.0807.17

Khan, S. (2010). *Strategies and spoken production on three oral communication task: A study of high and low proficiency EFL learners* [Doctoral Dissertation]. Spain: Universitat Autonoma de Barcelona.

Kothari, C. R. (2004). *Research Methodology: Methods and Techniques*. New Delhi: New Age International.

Kustati, M. (2012). The contribution of English students' speaking strategies and motivation on their speaking ability at Tarbiyah Faculty of Iain Imam Bonjol Padang. *Al-Ta Lim Journal, 19*(1), 9–16. doi: 10.15548/jt.v19i1.2. 362–369.

Lan, L. S. (1994). Fluency and accuracy in spoken English: Implications for classroom practice in a bilingual context. *The English Teacher Journal, 23*, 1–7.

Lan, V. T. N., & Lucas, G. R. I. (2015). The role of attitude, motivation, and language learning strategies in learning English as a Foreign language among Vietnamese college students in Ho Chi Minh city. *Asian Journal of English Language Studies, 3*, 1–26.

Liansari, V. (2016). Successful English learners in speaking English at SMAN 2 Surabaya. *Journal of English Educators Society, 1*(2), 115–122. doi: 10.21070/jees.v1i2.443

Lochana, M., & Deb, G. (2006). Task-based teaching: Learning English without tears. *Asian EFL Journal Quarterly, 8*, 104–164.

Ma, T. (2009) An empirical study on the comparison between strategies on oral English classes and writing English classes. *English Language Teaching, 2*, 39–45.

Martina, S., Syafryadin, S., & Salniwati, S. (2020). Compensation strategies in speaking activities for non-English department students: Poor and competent speakers. *Journal of English Educators Society, 5*(2), 7–14.

Mistar, J., & Umamah, A. (2014). Strategies of learning speaking skill by Indonesian learners of English and their contribution to speaking proficiency. *TEFLIN Journal – A Publication on the Teaching and Learning of English, 25*(2), 203–216. doi: 10.15639/teflinjournal.v25i2/203-216

Mistar, J., Zuhairi, A., & Umamah, A. (2014). Strategies of learning speaking skill by senior high school EFL learners in Indonesia. *Asian EFL journal press, 80*, 65–74.

Mohamed et al. (2020). Scaffolding the development of English language and communication skills of engineering students. *Universal Journal of Educational Research, 8*(5), 100–107.

Mohammadipour, M., & Rashid, S. M. (2015). The impact of task-based instruction program on fostering ESL learners' speaking ability: A cognitive approach. *Advances in Language and Literary Studies, 6*, 113–126. Retrieved from http://files.eric.ed.gov/fulltext/EJ1128408.pdf

Nayak, S. (2016). English for engineers: Meeting their needs of language learning. *International Journal of English Language, Literature and Humanities, 4*(3), 392–396.

Nguyen, T. B. H. (2013). English learning strategies of Vietnamese tertiary students. Published Doctoral Dissertation, University of Tasmania.

Panda, R. C., & Patnayak, K. K. (2017). English communication competence: Sine qua non for professional and personal development in engineering. *International Journal of English Research, 3*, 13–16. Press, 2007.

Rahman, N. A. A., & Maarof, N. (2016). The relationship between language learning strategies and students' motivation in learning English as A second language. *JTSS, 1*(1), 1–18.

Rani, S. M., & Jayachandran, J. (2014). Need for inclusive ESL curriculum incorporating necessities, lacks and wants. *Journal of ELT and Poetry, 2*(4), 424–431.

Ravi, S., & Sethuraman, M. (2019). *Developing speaking proficiency using OCT.* Germany: Lambert Academic Publishing Limited.

Safari, M, & Fitriati, S. (2016). Learning strategies used by learners with different speaking performance for developing speaking ability. *English Education Journal, 6*(2), 87–101.

Salikin, H., Zulfiqar Bin-Tahir, S., & Emelia, C. (2017). The higher achiever students' strategies in English learning. *Modern Journal of Language Teaching Methods, 7*(11).

Sien, O., Pool, D., & Rodrigues, P. (2018). Choice of language learning strategies: A case study of proficient and less proficient EFL students in the development of speaking skills of an intensive English programme.

Thanghun, K. (2012). *Using task-based learning to develop English speaking sbility of Prathomsuksa 6 students at Piboon prachasan school* [Master's Thesis]. Retrieved from http://thesis.swu.ac.th/swuthesis/Tea_Eng_For_LanMA/Kesda_T. pdf

Utama, F., & Shabir, M. (2017). Students' strategies in learning speaking. *English Journal, 20*(2), 1–13.

Verma, Prachi & Bhattacharya, Saumya. (2015, August 07). How English-speaking skills are keeping Indian engineers away from their dream jobs. *The Economic Times.* Retrieved from https://economictimes.indiatimes.com/jobs/how-english-speaking-skills-are-keeping-indian-engineers-away-from-their-dreamjobs/articleshow/48384078.cms

Wael, A., Asnur, M. N. A., & Ibrahim, I. (2018). Exploring students' learning strategies in speaking performance. *International Journal of Language Education, 2*(1), 65–71. doi: 10.26858/ijole.v2i1.5238

Wilona, A., Ngadiman, A., & Palupi, M. (2010). The correlation between intrinsic motivation and speaking proficiency of the English department students. *Magister Scientiae,* 45–55. doi: 10.33508/mgs.v0i27.645

Yousif, A. S. B. (2017). *The effect of communicative task-based instruction on developing students' oral communication skills at Sudanese Universities* [Doctoral Dissertation]. Retrieved from http://repository.sustech.edu/bitstream/handle/123456789/18759/The%20Effect%20of%20Communicative%20Task.... pdf?sequence=1&isAllowed=y.

Zhang, Y. (2009). Reading to speak: Integrating oral communication skills. *English Teaching Forum,* 47, 32–34. Retrieved from http://files.eric.ed.gov/fulltext/EJ923446pdf.

Chapter 11

Teaching Speaking in Indian ESL Classrooms

R. Nandhini and D. Poorvadevi

Engineering students need to be competent in their technical skills as well as in their ability to communicate in English effectively. There has been an increasing demand from employers for effective English communication skills from their job applicants. In some instances, lack of communicative skills has been touted as the reason for rejecting an applicant. Communication in English has become a much-needed skill in workplace contexts, especially if the employee has to work in a multilingual setting. India, with its myriad cultures, traditions and languages, poses a unique problem to its residents.

English in India

A language actively used in one corner of India is not recognizable in another part of the country. Colonization and the impact of British rule in India has created the dependence of Indians from different states on using English as their link language. English for communication is approved in almost all walks of life, including government offices, educational institutions and other multinational companies. Although, there have been many efforts to introduce one language as the common language, there have also been vehement disagreements and disputes from southern states, particularly Tamil Nadu in installing Hindi as the national language. India as a nation was formed from smaller states, each having its own language, culture and tradition. Thus, nominating one language as the national language has always been a bone of contention in India. English, as the neutral outsider, continues to be the language for communication in social and workplace settings, connecting people from various states.

Teaching English in India

Although English enjoys the status of second language in India, there are inherent issues that affect its acquisition. There remains a divide in the language acquisition based on economic segregation, locale, or the schools

DOI: 10.4324/9781003268529-16

where the students are educated. The larger population in India studies in schools with medium of instruction in the vernacular language. It could be said that even schools with English as the medium of instruction use the local language not only for teaching sciences and mathematics but also for teaching English. Thus, English is also taught like a subject rather than a language without any inherent change to the methodology of teaching. There is a fundamental flaw in the manner of teaching and learning English. Most students are able to define grammar terms but not use it proactively in communicative contexts. This lack of connect between learning and practice is compounded by many issues. Firstly, students easily pass English exams at all levels as the testing practices do not conform to testing communicative ability, rather, the skill of memory and rote learning. Almost all assessments of English for primary and secondary level education are in written form and are graded accordingly. Inclusions from grammar are prescriptive in nature and do not relate to language in everyday use. When students with this learning system face a pedagogically different approach to language learning, their affective and inhibitive factors increase. There is an association of negative emotions with learning where English language is concerned, and this affects their learning trajectory. Lastly, learners believe that bonding with peers is not possible if conversations are in English and thus friendships and associations are built and strengthened through use of native tongues. This creates an alien environment for use of English and in many instances, any conversation in English is thus forced and stilted.

English in Engineering Colleges

Language teachers in engineering colleges face a lot of obstacles as they are responsible for creating competencies in their language learners that will enable and improve employment opportunities for their students. Even so, students exhibit little to no motivation in classrooms and do not respond well to interactions in the language classes. The importance of communicating effectively is felt only when recruitments drives are organized in their respective institutions. Thus, any motivation for empowering themselves with soft skills is external with a search for a quick-fix solution. English language teachers are placed in an untenable position as they have to show improvement in their students' language ability.

Selecting Material for Teaching English

There is a demand for appropriate teaching material that can be used effectively. Many English language teachers are trained in literature and thus do not possess the theoretical knowledge of teaching English as a second language. This is especially important when handling unique situations of language learning/teaching. There is also an increasing need for suitable teaching material for English language classes. Background knowledge in

ELT gives teachers the ability to adapt existing material to suit individual needs of their students. It is also true that creating material for teaching is a time-consuming process. Hence, there is a heavy reliance on existing course books or activity workbooks. Teaching material available for purchase fall into two categories – ELT-based books that are expensive and books that are economic and locally available. The issue arises when there is no concrete evidence of theoretical reliance in the designing of tasks and activities.

Communicative language classrooms are based on the use of language in social contexts and the language user's need to perform specific functions in a social construct. Thus, training is focussed on listening, speaking, reading and writing (LSRW) skills along with grammar and vocabulary. While many learners exhibit different competencies in each skill, a competent language user will have similar levels of proficiency in all the LSRW skills. The occurrence of jagged levels of competencies can be noted in all skills, particularly in speaking. For example, a language user with an urban background may have better speaking skills while one from a rural setting may not. This is one of the critical considerations in a heterogeneous learning context.

Teaching Speaking: Practice and Issues

Speaking is the most noticed among all language skills. Communication is enabled most effectively through speaking and this skill aids in the formation of first impressions. The importance of speaking, thus, cannot be undermined. This is also the first skill that enables employers to gain perspective of a job applicant's overall proficiency. Language learning is intrinsically a psychological process and the proficiency of any individual cannot be measured accurately. Language ability may be affected by emotional or behavioural disarrays. Thus, teaching a skill that is fundamentally a psychological manifestation of the human psyche is often tricky; more so, when there are factors that actively hinder language learning.

A strong foundation in teaching and learning pedagogies is necessary for the teacher to adapt to situations specific to his/her classroom. It is to be noted that many teachers working in engineering colleges may have limited knowledge in ELT and teaching pedagogies. Recent amendments to courses for teachers have seen the inclusion of teaching knowledge and learning behaviours, but this does not reflect on the knowledge base of existing faculty in English departments country wide. Universities and educational boards have started taking steps to include this important facet of teaching in faculty development programmes. It is questionable whether the intent inherent in these programmes cumulate to actual acquisition. Increased student strength, decreased motivation, taciturn learners, inflexible class hours, poor infrastructure and limited material add to the problems faced by the language teacher. Teaching specific language skills with the aim to build holistic development of the language becomes a challenge and often results in poor performance of students in recruitment selection processes.

Since most of the soft skills are tested through speaking, it is imperative that students become proficient in speaking and display their language competency through oral skills. As mentioned earlier, the onus is on the teacher to help develop language skills of the student through these skills. Although no skill can be taught or learnt in isolation, language, holistically, is learnt through each of these skills. For example, a listening activity can be used as an example of speaking task, writing task easily converted to a reading activity and so on. Taking all these important factors into consideration, this chapter examines the process of developing or adapting tasks to help teach speaking in engineering classes. The inclusion of such tasks and activities in classrooms hopes to envisage improvement in the overall development of the language proficiency of the student.

Testing Speaking: Practice and Issues

If teaching speaking skills in English is mired in difficulties, testing the same skill faces more issues with respect to the design, production and implementation of test tasks that fulfil the purpose of the test. In theory, speaking test tasks measure the speaking ability of a test taker and provides information regarding the competency exhibited during the test. Although, an accurate measure of a language learner's ability cannot be identified within the constraints of a test, it still provides an overall idea of the language user's ability. A test must be valid and reliable to provide accurate information on the aspect that is being tested. However, a language test observes the qualitative ability of the test taken which is then quantified in the form of scores and grades. This transition from qualitative to quantitative scoring needs a strong foundation in validity and reliability of the test. In engineering colleges of India, there is a disconnect between what is taught and what is tested. Compounded by the lack of knowledge of testing principles, the selection of test tasks is based on the previous years' choices. This unfortunately does not provide a clear picture of the learner's ability in using the language. While a student is able to pass his/her end of term examination, the same language learner is unable to fulfil social or academic functions using the same language. A paradigmatic change in the assessment pedagogy will bring about changes in the way language is taught in classrooms. It is an inevitable truth that tests decide the teaching methods used in classrooms. If the test is more representative of the real-time communication, then the teaching method will also follow suit.

Literature Review: The Theory of Practice

Theories of language and learning provide ELT teachers with the anchoring that is needed to create and use tasks effectively in teaching events. Initially, language teachers used one theory exclusively to teach language, but over the years, it has become widely accepted that each learning situation is unique and that no single theory can encompass all the learning needs of that particular situation. Thus, a successful combination of theories can help in the creation and use of tasks. They can also help in predicting the outcome of the teaching-learning process. Conversation analysis (CA), communicative language ability and task-based language teaching are some of the theories that can be used in language classrooms to provide the necessary background to create activities in an interactional classroom, where learning is communication driven and each task has conversation as its core feature.

Conversation Analysis

CA studies and interprets facets of social interaction present in a conversation. CA deals with the instances of turn-taking, sequencing and adjacency, system of repair and preference organization in a conversation. Thus, analysing conversation provides the ELT teacher with information about possible openings, pre-closings and closings. While CA may not directly assist in creating speaking tasks for immediate use, it helps in identifying possible patterns that are commonly used in conversation. It also helps in providing a template for the teacher to create or to adapt a task for teaching speaking. Lazarton (2002) mentions that the most important goal of CA is to identify recurring patterns of organization that are created by a number of speakers including patterns not commonly used in social interactions.

Communicative Language Ability

Communicative approaches to language teaching provided the much needed emphasis that language teaching also included the development of communicative competence rather than proficiency in language structures. Hymes (1972), in his theory of communicative approach mentions that language teaching must predominantly develop communicative competence. Following this, many ELT practitioners worked on developing theories that postulate communication as the key to language teaching and learning. Canale and Swain (1980) have suggested that there should be three competencies that combine to produce an overall communicative competence of the speaker – grammatical competence, sociolinguistic competence and strategic competence. This highlights the need to understand the non-linguistic and para-linguistic factors that influence and affect communication. This study highlighted the important distinction between competence and performance. Thus, what is provided as a sample of speech may not be the evidence of actual competence.

Following the work of Canale and Swain (1980), Bachman and Palmer (1996) have proposed a theory of communicative language ability consisting of two components: organizational competence and pragmatic competence. This theory combined the influence of linguistic abilities with other non-linguistic features that affect communication (Figure 11.1).

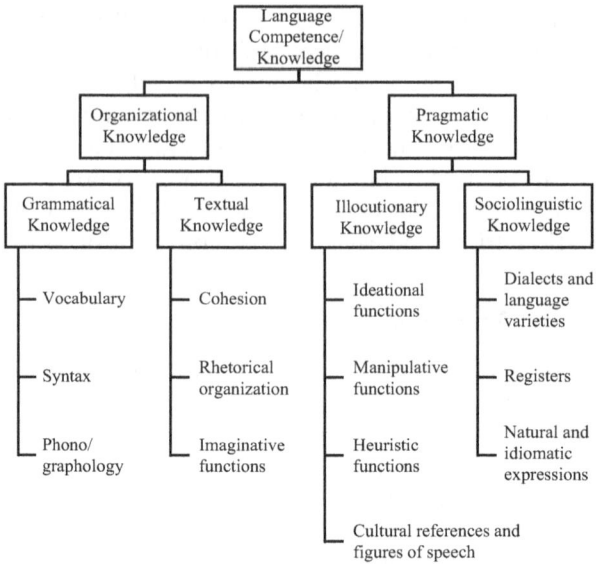

Figure 11.1 Theory of Communicative Language.

(Figure Based on Bachman & Palmer, 1996, p.68).

Task-Based Language Teaching (TBLT)

This theory is based on communicative approaches of language teaching and assumes that the 'task' is central to a language teaching and learning situation. As the name suggests, it proposes that tasks must be used as the learning tool in the language classroom. The choice of tasks as well as the nature of interaction lends credibility to TBLT as the use of tasks provides better contexts for learning. The Bangalore project undertaken by N.S. Prabhu in 1987 is a prime example of using tasks in the language classroom (as cited in Richards & Rodgers, 2009, p. 223).

Tasks can motivate learners to interact if they provide relevance and appropriacy to their learning needs. Such tasks provide the learners with a much-needed impetus to learn. Richards and Rodgers (2009) quote Skehan's suggestion that tasks can be designed to focus on specific aspects of language like fluency, accuracy, or grammar structures (p. 229).

Speaking Tasks

The following task types are used most commonly used in teaching speaking, but the tasks designed for each type will need to realize the purpose and the specifications intended for its use in a learning setting.

- Read aloud tasks
- Describing a picture/picture story

- Spoken essay
- Oral presentation
- Interview: controlled and open
- Information gap
- Role-play

(Weir, 1990)

Some of the tasks mentioned above may be adapted to the language class-room of engineering students. Each task type should be assessed on its purpose, use, design and ease of implementation.

Description of Tasks

Read-Aloud Tasks

These tasks are simple to recreate and as the name indicates, the students need to read extracts as part of the task. The challenge in this lies in the choice of reading material. In engineering contexts, technical manuals, guidelines, manuals and precaution information can be used as part of the reading material. Reading aloud provides the student with practice in pronunciation, tone, pitch and stress. Reading aloud also creates familiarity for a student with the texts and provides practice in engaging the anatomical part of sound creation.

Example: Read-aloud task

> A computer program is a collection of instructions to perform a specific task. For this, a computer program may need to store data, retrieve data, and perform computations on the data. A data structure is a named location that can be used to store and organize data. An algorithm is a collection of steps to solve a particular problem. Learning data structures and algorithms allow us to write efficient and optimized computer programs.

This text may be read aloud and then extended as a meaning generating activity. A read-aloud task is tightly controlled and does not allow for free interpretation by the student.

Describing a Story

The student will be asked to describe/talk about a picture. The difficulty in this task is based on the presence of prompts and the duration of the task. Similarly, the complexity of the task is directly related to picture selection (Figure 11.2).

- Use five sentences to describe the picture. You can use the following hints
- production line: Efficiency of automatic work-flow

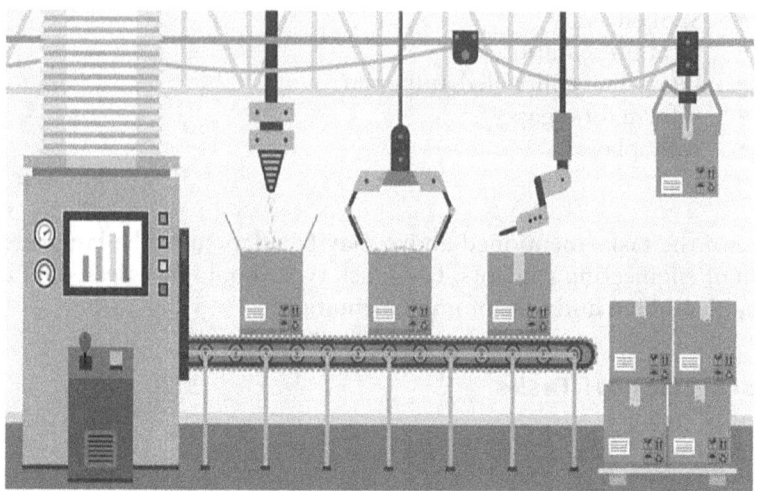

Figure 11.2 Illustration of Picture Description Task.
Source: Author.

If additional prompts are needed, the task can include hints in sentence form. This allows for more creative production of speech, yet restricted to the setting provided in the picture. Depending on the nature of the lesson, tasks may include pictures describing vacation spots, hobbies, landscapes, or other locations that are known to the students.

Oral Presentation

These tasks are more relevant to students as they are needed to talk on various topics related to their area of technical expertise. Also, job applicants need to be able to communicate effectively on a variety of topics in a small group as well as wider audiences. Oral presentation tasks should be designed in such a way that they enable extended speech, but shouldn't test knowledge regarding the topic or create strain for the student. Any task created should therefore include information about the topic in order to ease the process of presentation.

Example:
Describe a seminar you attended

- Who was the speaker?
- What was the topic?
- What did you learn?
- Was it useful to you?

Table 11.1 Illustration of Role-Play Task

Card 1	
Setting	Teacher's office
Student 1	You are late to class. You need to meet your teacher (S2).
	• Greet teacher and apologize for delay.
	• Explain that you had to undergo physiotherapy for 15 days.
	• Assure that you will be punctual after the therapy sessions end.
Card 2	
Setting	Teacher's office
Student 2	You are S1's teacher. You noticed him/her coming late to class frequently.
	• Ask S1 why he/she was delayed.
	• Be understanding, suggest the possibility of shifting the therapy sessions to evenings. Permit the student to arrive late for only these 15 days.
	• Advise the student to be punctual after the end of physiotherapy.

While this task provides guidelines, it also provides freedom for interpretation and the input needs to come from the student. This task requires the student to be adept at creating longer stretches of speech. The complexity of this task can be managed by changing the information given in the prompts.

Role-Play

Role-plays try to emulate a real-life situation in the classroom to provide students with the experience of participating in such instances. The role-play may involve one or two students depending on the existing language competence of the students. If a role-play is enacted between the teacher and a student, the task is guided and controlled by the teacher to a greater extent. Role-play tasks offer variety, scope and opportunities for students to use language in interaction. It can include a range of topics and can be used for varying durations (Table 11.1).

Task Development Cycle

Designing a task is a lengthy process that needs to fulfil the learning outcomes detailed out in a specific learning situation. A development cycle includes use, purpose, design, implementation and retrospection as part of its process. As with every development process, it is cyclical in nature and each part needs to be understood and completed for the task to be effective in the classroom. This chapter provides an illustration of the task development process for designing a speaking task that can be used for engineering students learning English.

Description of a Task Development Cycle

Use:	To use in English language classrooms of engineering students
Purpose:	To elicit information through guided and controlled approach
Level:	B1 – intermediate language user (please refer CEFR rating for level of language proficiency)
Design:	interview – controlled.

- Topics selected for task, one-to-one interaction with the teacher (facilitator).
- Provide opportunities for using a range of language structures – questions, expressing opinions – likes and dislikes, providing a summary of the discussion.
- Plan for openings and closings appropriate for the interview.

Duration:	3–4 minutes
Implementation:	Classroom/online (flipped classrooms).

- Provide opportunity for each student by splitting the interview sessions over a period of two weeks so every student gets an opportunity or use other resource persons to aid in the facilitation of the interview.
- It can be recorded for providing feedback. Students can also be assigned observation tasks to enable active listening in the classroom.
- This also provides additional learning opportunities.

Retrospection:	Effectiveness of the task must be analysed. Feedback from students and other ELT resource persons can be used to correct and fine tune the task.

Each task type needs to be carefully analysed for relevance in the classroom. The effectiveness of a task must be ascertained before it can be used again in a teaching-learning instance. Feedback from peers as well as the students provides information about the efficacy of the use of task for instruction in language classrooms.

CEFR Rating Scale

The Common European Framework for Reference (CEFR), published by the Council of Europe (2001), describes the ability of a language learner based on what the user can do using the language. This ability in separated into six levels. They are A1, A2, B1, B2, C1 and C2. A1is the basic language user while C2 is the mastery level of a proficient user. The 'can-do' statements developed by the Association of Language Testers in Europe (ALTE) describe possible communicative tasks that the language learner/user can perform at a specific level. The list of communicative language possibilities is comprehensive, and the can-do statements provide a concrete structure on which teaching and testing outcomes can be founded on. The efficiency of the CEFR levels and the corresponding 'can-do' statements has made it possible for the framework to be used in international assessments. The same can be adapted in the design of classroom teaching and testing scenarios (Figure 11.3).

Proficient User	**C2**	Can understand with ease virtually everything heard or read. Can summarise information from different spoken and written sources, reconstructing arguments and accounts in a coherent presentation. Can express him/herself spontaneously, very fluently and precisely, differentiating finer shades of meaning even in more complex situations.
	C1	Can understand a wide range of demanding, longer texts, and recognise implicit meaning. Can express him/herself fluently and spontaneously without much obvious searching for expressions. Can use language flexibly and effectively for social, academic and professional purposes. Can produce clear, well-structured, detailed text on complex subjects, showing controlled use of organisational patterns, connectors and cohesive devices.
Independent User	**B2**	Can understand the main ideas of complex text on both concrete and abstract topics, including technical discussions in his/her field of spcialisation. Can interact with a degree of fluency and spontaneity that makes regular interaction with native speakers quite possible without strain for either party. Can produce clear, detailed text on a wide range of subjects and explain a viewpoint on a topical issue giving the advantages and disadvantages of various options.
	B1	Can understand the main points of clear standard input on familiar matters regularly encountered in work, school, leisure, etc. Can deal with most situations likely to arise whilst travelling in an area where the language is spoken. Can produce simple connected text on topics, which are familiar, or of personal interest. Can describe experiences and events, dreams, hopes & ambitions and briefly give reasons and explanations for opinions and plans.
Basic User	**A2**	Can understand sentences and frequently used expressions related to areas of most immediate relevance (e.g. very basic personal and family information, shopping, local geography, employment). Can communicate in simple and routine tasks requiring a simple and direct exchange of information on familiar and routine matters. Can describe in simple terms aspects of his/her background, immediate environment and matters in areas of immediate need.
	A1	Can understand and use familiar everyday expressions and very basic pharses aimed at the satisfaction of needs of a concrete type. Can introduce him/herself and others and can ask and answer questions about personal details such as where he/she lives, people he/she knows and things he/she has. Can interact in a simple way provided the other person talks slowly and clearly and is prepared to help.

Figure 11.3 CEFR Global Scale.

Source: Authors Based on the Report from the Council of Europe (2001).

Conclusion

The language teacher may use existing tasks or modify them to suit the needs of the students. Activities may also be designed and developed from scratch. This gives the teacher the added credit of creating his/her own tasks for instruction and provides a much-needed variety in the classroom. Speaking is a skill that is rooted to the psychological framework of an individual and each individual perceives and acquires this skill differently. Degrees of competence can be observed in individuals and thus teaching to a heterogeneous classroom with mixed competencies provides a challenge to the teacher involved. Using interactional tasks may provide motivation and generate interest in the students, thereby enabling the learning process.

This chapter is an attempt at demystifying the process of developing tasks for teaching speaking of English in engineering colleges. By providing theoretical underpinnings to the methodology used in classrooms, ELT teachers can understand the nature of instruction needed to enable better learning in language classrooms. The chapter highlights the task development process by explaining how some tasks may be developed or adapted from existing models. Pre-existing knowledge of teaching strategies, material production and classroom management are some of the aspects that a language teacher must possess in order to make a positive impact on the teaching-learning process.

References

Bachman, L. F., & Palmer, A. S. (1996). *Language testing in practice: Designing and developing useful tests*. Oxford: Oxford University Press.

Canale, M., & Swain, M. (1980). Theoretical bases of communicative approaches to second language teaching and testing. *Applied Linguistics, 1*(1), 1–47. doi: 10.1093/applin/1.1.1

Council of Europe. (2001) *Global scale - Table 1 (CEFR 3.3): Common reference levels*. Retrieved from https://www.coe.int/en/web/common-european-framework-reference-languages/table-1-cefr-3.3-common-reference-levels-global-scale

Hymes, D. (1972). On communicative competence. In J. B. Pride, J. Homes, & H. Penguin (Eds.), *Sociolinguistics: Selected readings* (pp. 269–293). Harmondworth: Penguin.

Lazarton, A. (2002). *A qualitative approach to validation of oral language tests – Studies in language testing, 14*. Cambridge: Cambridge University Press.

Richards, J., & Rodgers, T. (2009). *Approaches and methods in language teaching*. Cambridge: Cambridge University Press.

Weir, C. J. (1990). *Communicative language testing*. UK: Prentice Hall International (UK) Ltd.

Chapter 12

Improving Reading Proficiency of ESL Learners

B. S. Prameela Priadersini and C. Harishree

Reading is a complex communicative process of receiving and interpreting the written word. The reader and the text are the two important components in reading. It is what happens in the mind of the reader when he goes through the printed letter. Reading is a visual task, recognition of words, reproducing the writer's meaning, a thinking process, and a step forward to personal development. In this line, reading has several definitions. Reading happens naturally as any other comprehensible aspect of existence. It is merely decoding, comprehending, and analyzing of words, and responding to it, but it needs continual practice. The real reading is the ability to understand the content and interpret it, and not merely reading the words without understanding them.

Reading is a psycholinguistic process that starts with a linguistic surface representation encoded by a writer and ends with meaning which the reader constructs in his mind. Thus, there is an essential interaction between language and thought in reading. The writer encodes thought as language and the reader decodes the language to thought. The reading process ranges from the simplest decoding of words to interpretative extensions beyond the author's message according to the experiential background of the reader (Ahuja & Ahuja, 1991). Reading process is not simply a matter of extracting information from the test, rather, it is one in which the reading activates a range of knowledge in the reader's mind that he or she uses and that in turn may be refined and extended by the new information supplied by the text. A key assumption is that whatever the readers perform during reading is not random but is the result of the reading process, whether successfully used or not. Reading is thus viewed as a kind of dialogue or interaction between the reader and the text. Reading process is a two-fold process, say a sensory process and a perceptual process. As a sensory process, reading is dependent on certain visual skills. It involves identification of symbols wherein eyes play an important role. The eyes look at the text and perform horizontal and vertical movements. The letter of the words is recognized, the words are identified, and the meaning are assigned to them. As a perceptual process, it refers to the interpretation of everything that one reads.

DOI: 10.4324/9781003268529-17

The mind processes the meaning in relation to its earlier knowledge, interprets, and understands it. The reading process thus involves identification of the visual symbols and association of meaning with these symbols.

For the students of engineering, reading is very much important to achieve success in their career because reading can improve their vocabulary thereby providing them confidence to communicate well and get a good placement. It can also help to improve the other three language skills namely listening, speaking, and writing. The students of engineering need to communicate their technological ideas to others for their career prospects. In the 21st Century workplace environment, English will be their main medium of communication and they will be interacting with people who have different mother tongues in the global workplace. They may have to read materials of various types and they require good command over English language and its vocabulary. Reading improves their vocabulary which paves way for good communication. Tsai and Chang (2014) have stated that English reading comprehensions is very crucial for their academic and future workplace. Further, the authors have stated "to absorb fundamental knowledge in textbooks and deepen professional knowledge in their workplace, it is essential for engineering students to read in English fluently" (p. 1). Reading also familiarizes them with the usage of the language at particular context which helps them know the structure of English.

According to the IRA (International Reading Association) report, excellent reading ability is essential for the success of the students ("Reading Today", 2008). The critical role of teacher in fostering students' success concerning reading has been accepted by educationists. Hagley (2017) has stated, "improving reading in a foreign language is both difficult and time consuming. Competition for students' time is fierce and engineering faculty at universities are often wary of anything that takes too much time away from their students" (p. 204). In this line, the strategies that can be taught to the students to improve reading skills are under debate. Rajasekar (2006) has demonstrated the feasibility of teaching a few strategies to students who have not been familiar with using such strategies in their ESL reading. However, there has not been more studies exploring the students of engineering vis-à-vis their reading skills especially in English as a Second Language (ESL). As a result, this creates a necessity to undertake this study and teach them certain strategies to improve their reading skills. In this line, the present study attempts to make the students learn reading strategies which improve their reading speed, comprehension level and retention level eventually during their practices for a period of time.

Purpose of Reading

People read for different purposes. It may be for pleasure, to know things, to gain knowledge, and to relax. Hathaway (1929) has identified 1620 purposes of reading which have been classified under nine major heading namely, to gain meaning, to gain information, to guide activity, to find values, for social motives (that is, to influence or entertain others), to organize, to solve problems, to remember, and to enjoy. Commenting on the results of various studies, Gray (1965) has pointed out three conclusions as follows:

1. Reading is used for a surprisingly wide variety of purposes.
2. The purpose of reading varies from on curricular field of study to that of another field.
3. The purpose changes from one level of scholastic advancement to another.

Purposes of reading in classrooms are equally revealing, some of these which have been emphasized repeatedly are as follows:

1. To find answers to specific questions
2. To determine the author's aim and purpose
3. To follow a sequence of related events
4. To find the central thought of a selection/text
5. To find the most important points and their supporting details
6. To select facts which relate to a problem
7. To judge the validity of statements
8. To find fact supporting a point of view
9. To draw valid conclusions from materials read
10. To discover problems for additional study to remember what is read (Ahuja & Ahuja, 1991)

Accordingly, the purpose of reading is to gain knowledge and also relax during certain serious situations. Through reading, one can come to know various events that happen throughout the world, improvements in various fields, recent trends, requirements of job market, etc.

Theories of Reading

There are three theories/models on how reading occurs.

Bottom-Up Theory

The bottom-up theory argues that the reader starts with the individual letter followed by how the letter are combined to form words, how words are combined to form phrases, and finally how phrases into sentences. Thus, the reader starts from the smallest unit and slowly moves through higher states of processing. The process of constructing the text from these small units becomes so automatic that readers are not aware how it happens (Eskey, 1988; Stanovich, 1990). This model emphasizes what is on the printed page and particularly on the visual information and not on what the reader brings to the text. In other words, 'bottom-up' process is activated by the incoming data. Hence, it is known as data-driven approach.

Top-Down Theory

In this model, the reader brings a great deal of knowledge expectations, assumptions, and questions to the text and is given a basic understanding of the vocabulary; he continues to read as long as the text confirms his expectations (Goodman, 1967). This theory emphasizes the role of stored information rather than perceptual one. Ghanta (2019) has stated, "top-down reading theory underscores from the thoughts to text of readers who study their methodology centering on the text of the content by contradicting the real substance of the text" (p. 520). The reader tends to use minimal cues from the page. He fits the meaning into knowledge he already possesses and then checks back when new information appears. His cognitive and language capability play the most significant role in his ability to comprehend the text. Hence, it is known as 'conceptually driven' approach.

Interactive Model

An analysis of the top-down and bottom-up components of the reading process shows that neither of them is complete in itself. So as an alternative model, Rumelhart (1977) has developed an interactive model of reading. This strikes a balance between 'data-driven' and 'conceptually-driven' approaches, and so, it is endorsed by most of the current researchers (Barnett, 1989). This process moves both bottom-up and top-down, depending on the type of text as well as on the readers' background knowledge, language proficiency level, motivation, strategy, and culturally determined beliefs about reading (Carrell, Devine & Eskey, 1988; Barnett, 1989).

ESL Reading

ESL reading theory has changed dramatically from bottom-up model to interactive process. According to Widdowson (1979), reading is a process of combining textual information with the information a reader brings to a text. It is viewed as a dialogue between the reader and the text. Understanding reading is an interaction that occurs between the reader and the text. It is an interpretative process. This perspective has evolved out of the reading research of Goodman (1970, 1976) and Smith (1982), as well as from the development of schema theory. Goodman and others have had an opinion that reading is primarily concept driven. This has been referred to as top-down approach to reading (Gough, 1984). Goodman's approach has led to extensive research on how conceptual knowledge, inference, and background information affect the reading process, particularly in ESL reading. This approach to reading has had a powerful impact on ESL reading theory and practice.

Classification of Language Learning Strategies

Learning by strategies is a foundation for lifelong learning. Strategy is a series of well-planned steps for achieving an aim. In a learning situation, a strategy is a method or way that can be learned and developed to cope with various learning tasks. They are the tool for the self-directed involvement necessary for developing communication ability. Strategy is a method which helps the learners to successfully completing a specific task. Strategic learning is a key to students' success in any endeavor. Accordingly, learning strategies are means through which a learner is able to acquire the knowledge of second language, and as a result, enable them to read and comprehend well. The learner makes use of certain actions to make his learning easier and enjoyable to acquire his efficiency. Learning strategies play an important role in second-language acquisition (SLA) and this has been highlighted by numerous studies. Dadour and Robbins (1996), Leaver and Oxford (1996), and Oxford (1996) have demonstrated that when strategy instruction is tailored to suit specific contents, and individual learner needs, it can be an effective method for speeding up the process of learning a second language.

In her taxonomy, Oxford (1996) distinguishes between direct and indirect language learning strategies (LLS). While direct strategies facilitate comprehension, indirect strategies support language learning indirectly acting upon the target language as "a strategy is a sequence of activities, not a single event and learners may have acquired some of the sequence, but not all" (Garner, Macready & Wagoner, 1984, p. 123). Chamot and O'Mallety (1994) have introduced three inter-related function-based strategy clusters, cognitive (used to accomplishing a specific cognitive task during reading, such as inference and word-part analysis) and metacognitive (used when interacting co-operatively with others during reading such as seeking outside assistance). Thus, the strategies are broadly divided into

1. Cognitive strategies
2. Metacognitive strategies
3. Social strategies
4. Affective strategies

These strategies can be taught to the students of engineering depending upon their level of understanding and the content of the reading material.

Reading Strategies

Reading strategies are techniques and methods readers use to make their reading successful. These methods include how to conceive a task, what textual cues they attend to, how readers make sense of what they read, and what they do when they do not understand. Reading is an interactive top-down and bottom-up process. The reader forms the meaning of the text through interaction of a variety of their mental processes to work at different levels such as using the bottom-up process to identify the meaning and grammatical category of word, sentence, syntax, and text details (Aebersold & Field, 1997). The reading process also gives rise to the issue of reading strategies. ESL and EFL learners usually employ a number of LLS during their reading process. Those strategies involve cognitive, metacognitive, compensation, memory, affective, and social strategies. (Chamot & O'Mallety, 1994; Crandall, Jaramillo,, Olsen, & Peyton, 2002; O'Malley & Chamot, 1990; Oxford, 1990).

Reading strategies are considered as essential aspect of teaching English as a foreign or second language. Indeed, many English text books for ESL/EFL learners emphasize focused strategy instruction in reading. Some of the reading strategies include skimming and scanning, summarizing information, making guesses, prediction, making inferences, understanding words or phrases, and making notes. Reading strategies help the second language learners to read the text and understand it using certain techniques. These methods normally facilitate the reading speed and comprehension, which create an interest among the readers.

As far as the reading process is concerned, the mental activities that readers use in order to construct meaning from a text are often referred to as reading strategies or reading skills. Strategies are primarily classified as knowledge-based strategies which are used in bottom-up processing. Global comprehension of a text requires knowledge-based strategies like previewing, activating background knowledge, whereas local comprehension requires language-based strategies like finding contextual clues, identifying grammatical structurers, etc. Two broad distinctions are common in these diverse categorizations. One distinguishing between cognitive and metacognitive strategies and the other differentiation local and global information-processing strategies. Studies indicate that these distinctions are useful in identifying the difference in strategy use among readers. Every reader encounters problem while reading, but what distinguishes a successful reader from others is his ability to use the correct strategy and comprehend better. The reader must be able to sense the nature of the problem, and give its possible solutions.

Strategic Reading

Strategic reading enables the readers to elaborate, organize, and evaluate the information derived from text. It enhances learning throughout the curriculum. It reflects metacognition and motivation because readers need to have both the knowledge and disposition to use strategies. It has been found for instance that first-language readers use global strategies to a greater extent than local strategies, whereas novice readers rely more heavily on local strategies (Myers & Paris, 1978; Paris & Jacobs, 1984). A similar tendency is noted in studies involving high- and low-proficiency second-language readers. Strategic reading cannot be accomplished without the readers' desire and intent to read more efficiently. The acquisition of strategic reading depends on the corresponding development of cognitive and metacognitive capabilities. Metacognition is best defined as thinking about thinking. As second-language readers actively monitor their comprehension process during reading, they will select strategies that will assist in arriving at the meaning. Metacognitive awareness of the reading process is perhaps one of the most important skills second-language readers can use while reading. This indicates that they are able to verify the strategies they are using. Moreover, recent research comparing the effectiveness of cognitive and metacognitive strategies training shows that implicit instruction on cognitive strategies, yields small, short-term improvements in reading performance, whereas training on metacognitive strategies results in more stable long-term comprehension gains.

Reading Strategies Employed in the Study

Grabe (1991) provides a caution, "effective strategy training is not a simple or easy matter" (p. 393). He points out that the duration of training, clarity of training procedures, student responsibility, and strategy transfer are variables that influence strategy training results. Several studies have shown that fluent word reading helps comprehension (Bell & Perfetti, 1994; Fuchs et.al., 2001). Struggling readers often experience significant improvement in comprehension, when taught reading strategies (Shearer, Ruddell & Vogt, 2001; Vogt & Nagano, 2003). Globerson, Weinstein, and Sharabany (1985) have found that when learners are focused on the activity of learning (e.g., through metacognitive training) their engagement and comprehension are enhanced. Lack of reading fluency is a reliable predictor of reading comprehension problems (Stanovich, 1991).

A cognitive understanding of what should be done is not enough to guarantee success while reading. The reader must also understand how to apply the use of a given strategy. In the early stages of reading comprehension, open discussions with the reader will be the best method to verify whether the strategy is being used appropriately. The use of verbal think-aloud protocols can facilitate the evaluation of the strategy. The researcher has adopted a two-pronged approach regarding strategies to improve the reading proficiency of the participants, one was prior to actual reading and the other was while reading. In the selection of suitable strategies for ESL reading, the researcher made use of the strategies suggested by Oxford (1990), Clarke (1979), Barnett (1989), and Anderson, Bachman, Perkins, and Cohen (1991). For instance, Anderson et al. (1991) have listed 24 strategies into three different groups cognitive reading strategies (thinking), metacognitive reading strategies (thinking about one's own your thinking/planning) and compensating reading strategies. Based on these studies, the current study has employed the following strategy to improve the students' reading proficiency.

1. Creating awareness among the students regarding the reading speed
2. Giving contextual clues
3. Guessing the meaning of unfamiliar words
4. Giving meaning for difficult words
5. Predicting the next sentence
6. Repeated reading
7. Re-reading
8. Prior discussion of the passage
9. Identifying the central idea of the passage
10. Developing sight vocabulary
11. Creating awareness of metacognitive strategies

Reading Efficiency

Efficient reading implies clear comprehension of the communication presented in print or written in a short time. Comprehension and rate of reading are the two main factors leading toward efficiency in reading. Reading efficiency is not a single skill but a whole repertoire of skills. In a study that explored the development of fluent and automatic reading, Kuhn et al. (2006) have reported that as students read more and more challenging texts devoting more time to read, they improve both in word reading and reading comprehension. As efficient reader always has a variety of reading speeds, it makes use of at least four types of reading rate:

1. Skimming rate
2. Rapid rate
3. Normal rate
4. Careful rate

Developing Efficient Reading Rate

Development of efficient reading rate is an important objective at all stages. Researchers have shown that the students who read rapidly and comprehend well have an advantage over the slow readers. Sometimes students read very fast without understanding the meaning, so they may get only incomplete idea of the passage of what they read, and in due course, it would become their habit, so any material should be read as rapidly as it can be comprehended while reading for a particular purpose (Bond & Tinker, 1957). Efficiency in reading depends on the reader's motivational readiness. Lack of interest is an important cause of inefficient reading. To be an efficient reader, the student must have the intention to learn. Interest serves as internal motivation. Emans and Patyk (1967) note that interests are influenced by the nature of the topic and the motive of the reader. Reading depends upon the nature of the topic, that is, how it could create interest for the reader to read and it also depends upon the motive of the reader, whether he reads for getting information or to identify certain things or for the aesthetic sense of the context matter or as a source of recreation.

Reading Speed

The most efficient reader is the one who can vary his speed appropriately in terms of the reading task that he is engaged. Reading speed is the time taken by a reader to read a passage and also to comprehend on it. Reading speed never remains constant. It differs from individual to individual. It also depends upon the material they read and the reason why they read a particular material. Merely reading the words without understanding cannot be defined as reading speed.

Increasing Reading Rate

Improving students' word recognition efficiency and helping readers develop greater sensitivity to the syntactic nature of the text will result in more efficient reading and improved reading rate. Understanding the concept of automaticity is a great value to a second-language teacher. Increasing students' reading rates makes them develop greater cognitive capacity to improve their comprehension level. This will allow students to move through the heavy assignments and studies. Often students are trying to focus on too many reading tasks at the same time and do not comprehend what they are reading. In addition, their reading rate is very slow which also results in not understanding what they have read.

Dubin and Bycina (1991) have stated, "a rate of 200 words per minute would appear to be the absolute minimum in order to read with full comprehension" (p. 190). Jensen (1986) recommends that second-language readers seek to "approximate native speaker reading rates and comprehension levels in order to keep up with classmates" (p. 106). She suggests that 300 words per minute is the optimal rate. This rate is supported by Nuttall (1982) who states, "for an L1 speaker of English of about average education and intelligence reading rate is about 300 wpm. The range among L1 speakers is very great; rates of up to 800 wpm and down to 140 wpm are not uncommon" (p. 36).

Reading Comprehension

Comprehension is the very heart of the reading act. It is assumed that comprehension would be natural consequence of decoding alone. But it is not so, in fact, the word comprehension itself is still to be invented in some of the languages of the world. In a few countries, the word: comprehension, and the expression: reading comprehension, have been used by educators, but readers in those countries still do not comprehend what they read, so well, as they should. Yoakam (1951) has described comprehension as comprehending reading matter which involves the correct association of meanings with word symbols, the evaluation of meaning which are suggested in content, the selection of the correct meaning, the organization of ideas as they are read, the retention of these ideas, and their use in some present or future activity. Comprehension is a complex process and it has been understood and explained in a number of ways. The RAND Reading study group (2002) has stated that comprehension is "the process of simultaneously extracting and constructing meaning through interaction and involvement with written language". Duke (2003) has added 'navigation' and 'critique' to her definition because she believed that readers actually move through the text finding their way evaluating the accuracy of the test to see if it fits their personal agenda and finally arriving at a self-selected location. Comprehension means reading a passage and understanding the usage of vocabulary, its literal meaning, and the implied meaning. Learners use various methods and techniques involved in grasping mentally understanding ideas/facts and comprehending the material what they have read. The understanding ability of the learner constitutes to comprehending factor and that is what constitutes comprehension.

Retention

Retention is the ability to retain facts and figures in memory which is a necessary requirement for engineers. They have to remember the matter that they have read. This is related to their comprehension level and memory capacity. Retention can be made possible by repetition and reinforcement. The readers normally read for understanding. As they read regularly their reading speed increases and when they read and recite or retell what is in that passage, they may be able to comprehend better and when they repeat it and review it again and again it would get into their mind and the content would be retained in the mind. Many standards of learning require students retain information what Jensen (1998) called semantic memory. The repeated content eventually becomes reflexive or automatic. Ideally, it enters the long-term memory – retained for performance and assessment. In this line, the equation for retention is as follows:

Reading + Recitation + Repetition + Review = Retention

Research Questions

1. Does reading speed have impact on reading comprehension level of the students?
2. Does reading strategies help in improving the retention capacity of the students?

Methodology

This study was conducted at Trichy Engineering College, Tamil Nadu, India, between December 2006 and March 2007.

Participants

Students from four branches of engineering namely Computer Science and Engineering (CSE), Electronics and Communication Engineering (ECE), Information Technology (IT), and Mechanical Engineering (Mech. Engg.) had participated in this study. Most of them had regional (like Tamil, Malayalam, and Telugu) medium of instruction during their schooling. The participants were from rural villages and semi-urban towns. About 36 students from all the four branches who studied in schools with Tamil medium of instruction were taken as participants for this experimental study. All the students were from rural areas, and more than 70% of them were first-generation learners.

As the study was conducted after college hours, only boys were chosen for this study considering the transport facility and safety precautions of the girl students. The participants were also been selected based on their reading habit in their first language, on their interest in becoming engineers, and awareness of importance of English for their future career. Additionally, the participants were chosen based on the criteria of unawareness of English reading that could be a vital factor for their career success.

Tools Used for the Study

Two questionnaires were administered for the study. The questionnaire elicited the responses on the participants' demographic details like location, schooling, etc., and on analyzing their reading skills. After the structured reading course for ten days, newspapers were given to the participants and they were asked to read and note down what they had read. Two passages in the regional language were given first to know their reading skills in their first language and ten passages in English of different length and varied levels of difficulty were selected to be given in the course of study. Ten passages consisted of three stories, two passages on science subject and three passages on subject of general interest, one passage on an industrial subject, and one on an adventure. The passages were given to the students in the order of their difficulty level. They started with simple passages and then proceeded to complex passages. The difficulty level depended on the sentence structure, organization of text, vocabulary, and the content of the passage. All the ten passages were chosen for their exploitability aspect also. Whether each passage lent itself to the suitability of teaching a specific strategy was also considered while choosing them.

A diary was maintained to note down the significant activities, queries, changes, and practices that the learners had during the course of the structured study. The participants' self-learning figures such as reading speed, comprehension, etc., on the passages they read during vacation was also entered in the diary. Further the retention level of the participants at the end of the course was also recorded in the diary. This was found out by questioning the participants individually. The oral interaction happened during the study was entered in the diary sequentially.

Testing Procedure

Most of the English text books would have a major section on assessing reading which gives a detailed explanation regarding various kinds of test items, instruction to follow on how to construct items. Some of the common tests used to test reading comprehension are

1. Vocabulary test
2. Cloze test
3. Completion task
4. Short answer questions and open-ended questions
5. Contextualized or authentic tasks (e.g., discussion)

In this study, the student's global and local comprehension ability, and critical and analytical thinking abilities were tested through the following types of questions:

1. Questions on vocabulary
2. Global and local comprehension
3. Inferential question
4. Critical questions
5. Analytical questions
6. Descriptive questions

All these types of questions were not asked for each of the passage. Depending on the exploitability of the text particular types of questions were asked.

Results and Discussions

The analysis of the first questionnaire revealed that the students had positive attitude toward reading and they had the habit of reading in their mother tongue on their own-interest. They were not aware of any strategies they followed while reading but they read some matter in their mother tongue regularly. Students who have judged themselves as average readers have performed well compared to the students who have judged themselves as good and excellent readers.

The analysis of the second questionnaire revealed that the students understood the importance of reading in English and how to follow certain strategies to improve their reading speed, comprehension level and retention level. They developed a positive attitude toward reading in English and realized the importance of it in improving the other three basic skills. They came to a strong conclusion that reading is a vital factor which can help them achieve their career goal.

Self-Assessment and Actual Performance – An Analysis

In the first questionnaire, a question was given to the students to judge themselves (self-assessment) as excellent, good, average, and poor reader. Most of them assessed themselves as average readers, two as good, two as excellent readers, and one as a poor reader. Table 12.1 shows the correlation between the self-assessment and actual achievement of these students. These results indicate that they were not aware of their own competence.

It is evident from the Table 12.1, students who have judged themselves as excellent readers, good readers, and poor reader have got more or less the same comprehension score and their reading ability is also nearly the same; some of the readers who have judged themselves as average readers have performed well than the excellent and good readers, which shows that they are not able to assess themselves. Further, the comprehension score of the excellent reader increased from 20% to 100% and that of good reader from 26.66% to 100% and 40% to 80% and that of poor reader from 13.33% to 100%. Though the poor reader has got a score of 13.33% in the first passage, his continuous involvement and the use of strategies has made him achieve 100% score in the last passage which explicitly shows that any student motivated and given regular practice can improve his reading speed and score.

Table 12.1 Self-Assessment vs Actual Performance – An Analysis

Sl. No.	Roll No.	Self-Assessment	1st Passage (Ad. Of sea) 440 words		10th passage (India as a hub for innovation) 1350 words	
			Wpm	%	Wpm	%
1	2k6415	Excellent	40	20.00	147	100
2	2k6436	Excellent	62	20.00	188	100
3	2k6212	Good	44	26.66	186	100
4	2k6530	Good	62	40.00	188	80
5	2k6213	Poor	62	13.33	167	100

Improvement of Reading Speed and Comprehension Level

Kumari (2016) has stated, "fluency in reading is defined as the ability to read with proper speed, pronunciation and pause/punctuation" (p. 74). In this line, the data of the time taken by 36 participants to read the first passage and the last passage and the corresponding comprehension scores are recorded and provided in Table 12.2. This is the time taken to read the

Table 12.2 Reading Speed and Comprehension: 1st Passage and Last Passage

Sl. No.	Roll No.	1st Passage			Last Passage		
		Time taken to read in mts.	Wpm	Comprehension %	Time taken to read in mts.	Wpm	Comprehension %
1	2k6209	5.00	88	33.33	9.20	146	70
2	2k6211	6.00	73	6.60	8.15	166	60
3	2k6212	10.05	44	26.66	7.25	186	100
4	2k6213	7.05	62	13.33	8.10	167	100
5	2k6406	6.20	71	33.33	8.45	160	100
6	2k6409	8.25	53	46.66	9.10	148	100
7	2k6415	11.00	40	40.00	9.20	147	100
8	2k6423	13.00	34	33.33	8.10	167	90
9	2k6432	15.00	29	46.66	9.10	148	100
10	2k6434	8.30	53	33.33	10.00	135	100
11	2k6436	7.10	62	20.00	7.20	188	100
12	2k6438	8.20	54	13.33	8.10	167	40
13	2k6441	9.10	48	6.60	9.20	147	90
14	2k6508	7.05	62	20.00	9.40	144	80
15	2k6512	10.20	43	33.33	8.20	165	100
16	2k6515	9.20	48	40.00	9.45	143	80
17	2k6530	7.15	62	40.00	7.20	188	80
18	2k6535	8.00	55	13.33	8.15	166	60
19	2k6538	6.50	68	33.33	6.00	225	90
20	2k6543	7.25	61	13.33	9.10	148	90
21	2k6545	5.50	80	26.16	8.20	165	100
22	2k6548	9.20	48	20.00	7.20	188	80
23	2k6553	8.50	52	20.00	7.45	181	90
24	2k6555	10.00	44	40.00	8.00	169	100
25	2k6557	7.50	59	26.66	7.45	181	100
26	2k6705	11.00	40	20.00	8.00	169	70
27	2k6715	9.50	46	33.33	7.00	193	80
28	2k6719	6.50	68	33.33	9.15	148	90
29	2k6720	10.10	44	26.66	9.45	142	100
30	2k6723	9.40	47	33.33	7.00	193	90
31	2k6727	10.25	43	53.33	8.25	164	90
32	2k6729	11.00	40	40.00	7.00	193	60
33	2k6739	6.25	70	26.66	7.20	188	80
34	2k6740	5.40	81	46.66	8.05	168	70
35	2k6745	9.25	48	33.33	8.10	167	100
36	2k6746	11.20	39	26.66	8.30	163	90

passage for the first time. This data shows that the reading speed of each student has increased.

Student No. 9 (Roll. No. 2K6432) in Table 12.2 had taken the maximum time of 15 minutes to read the first passage which has 440 words; the same student has read the 1350 words in the last passage in 9.20 seconds. His reading speed has increased from 29 words per minute in the first passage to 148 wpm in the last passage. His comprehension score also has increased from 46.66% in the first passage to 100% in the last passage. Student No. 1 (Roll No. 2K6209) in Table 12.2 has taken the minimum time of five minutes to read the first passage and his reading speed was 88 wpm. The same student took 9.20 minutes to read the last passage at a reading speed of 146 wpm. His comprehension score also has improved from 33.33% in the first passage to 70% in the last passage. Among these two students, student No. 9 has scored better in the comprehension of the two passages than student No. 1. Though student No. 1 has revealed a high wpm in the first passage, he could not reach the reading speed of the student No. 9 in the last passage and also his comprehension level.

The overall analysis shows that there was a steady increase in the reading speed when compared from the first passage to the last passage. The time taken to read the passage decreased slowly and reading rate increased steadily. On the other hand, it is inferred that the comprehension level increased when compared from the first passage to the last passage. The students were able to write answers correctly even for inferential and critical questions in the last passage which they were not able to in the beginning of the study. So, the reading speed has significant positive impact on the comprehension level of the students.

Impact of Reading Speed on Comprehension Level

In spite of teaching all strategies, reading comprehension appears to depend on the individuals' mental application at the time of reading. Too much of confidence seems to have brought down the level of comprehension. The feeling that they know reading has pulled them down. Table 12.3 shows that there are students who have taken much time to read and scored better marks than those who have taken less time to read. This finding makes one wonder if fast reading really improves comprehension level. Fast reading without mental application may not enable good comprehension.

From Table 12.3, it can be realized that reading comprehension is a complex issue. Participants who read at a moderate speed have comprehended better and scored more marks than those who have read quickly. Although there is an improvement in the reading speed from passage one to passage ten among all the participants, it has not been linear or uniform. All the participants did acknowledge the use of strategies while reading the passages. (This was ascertained while interacting with them during the course of the study.) But they have been unable to pinpoint any precise or single

Table 12.3 Reading Speed and Comprehension – A Paradox

Sl. No.	Name of the Passage	Roll No.	Maximum Time	Score	Roll No.	Minimum Time	Score
1.	Adventure of sea	2k6432	15.00	46.66	2k6209	5.00	33.33
2.	Wise owl	2k6432	14.25	36.36	2k6209	5.10	45.45
3.	The feather	2k6432	12.00	33.33	2k6545	3.00	25.00
4.	Team work	2k6508	5.50	71.42	2k6538	3.00	78.57
		2k6727	5.50	71.42	2k6538		
		2k6543	5.50	92.85	2k6553		
5.	Sleep	2k6727	5.20	80.00	2k6723	3.30	60.00
6.	Lemon Tree	2k6543	10.06	50.00	2k6436	3.15	60.00
7.	Industrial needs	2k6530	9.20	87.5	2k6723	5.00	56.25
8.	Street kids	2k6211	9.50	37.5	2k6436	4.00	56.25
					2k6723	4.00	37.50
9.	Qualities of a good manager	2k6720	8.10	77.77	2k6740	3.50	55.55
10.	India as a hub for innovation	2k6720	9.45	100	2k6538	6.00	90.00
		2k6515	9.45	48.25			

strategy in a particular context. This is very much similar to what they had done while reading in their first language.

So, though the use of strategies has beneficial outcome, pinpointing the impact of individual strategies in improving reading speed/comprehension appears difficult. What emerges from the above analysis is the role of the mind in comprehending, where there is concentrated mental application, reading and comprehension seems to happen effectively, when there are digressions, the same is affected. It can be summarized as that reading speed has played crucial role in comprehension level of the students. Further, these participants have developed a flair for reading in English on the teaching of strategies and that has helped them to overcome their lack of self-confidence in reading English.

Retention Level

This study has attempted to find out the retention level of the passages by the participants. No specific/conscious attempt has been made to help the participants retain what they had read over a period of two weeks. The retention level of the participants has improved because when they have been asked to write and speak about the passages, they have read 15 days before, most of them have been able to remember the contents and give relevant details of the eight passages they have read. It has been exhibited in Table 12.4.

Table 12.4 Retention Level of the Students

Passage No.	Number of students who remembered
Adventure of sea	36
Wise owl	20
Feather	17
Team work	2
Sleep	4
Lemon tree	28
Industrial needs	34
Street kids	30

Table 12.4 gives a clear picture of the retention level of students for all the eight passages. All the students were able to remember the first passage since it was about an adventure and moreover it was the first passage given to them in the reading course. For all the students, it was their first experience where they were given a separate hour for reading and were asked to read a different material apart from their regular course of study. So, their thrill and enthusiasm increased. Moreover, the first passage was on discovery of Vasco da Gama and Columbus, which they would have studied in their school, so they would have had some background knowledge about it. This would have helped them to retain the content of the passage as nearly 50% of the students were able to give the gist of the stories. Since they were stories, they were able to remember, recall them, and say them without difficulty.

More than 90% of them were able to give the content of the passage 'Industrial needs' which was discussed before it was given to them to read during the course of their study. Since it was discussed, they were able to retain the key points and they also remembered the central idea of the passage. They were also able to give the information content of it. The two passages 'Team work' and 'Sleep' were not remembered by all because of their complexity in their vocabulary and structural organization. The passages were scientific and the content was difficult to understand; apart from that, they had many unfamiliar and difficult words which the students found difficult to guess the meaning while they were reading. Another possible reason for this may be the timing of the two passages when they were given to the participants. They were given in day 4 and day 5 of the study. After that, two days were holidays and on the eighth day passage 6 was given for reading after a general talk on motivating them to read. So, they might have forgotten them as the passages might not have created an interest in them.

The passage which gave them the main idea very easily was 'Street Kids'. Nearly 30 students were able to say about 'Street Kids' which talks about two social service organizations. Since they were able to paraphrase it and understand the main idea, they were able to retain it and recall it when needed.

Impact of Strategies

The objectives of this study are to improve the comprehension level of ESL learners, increase their reading speed and also their retention of what they had read. This was supposed to be achieved through the use of certain reading strategies. The strategies taught to the students seem to have had an impact on their reading habit. In the beginning, the students participating in this study were not confident of their reading of texts in English. But after this course, they had developed an interest in reading and most of them were practicing this skill.

They seem to be using the strategies in a multi-pronged manner. Interaction with them revealed that they were in the habit of guessing the word meaning from the context after this course. In this, they were using the contextual clues. By more reading, they had learned to predict the matter that will follow what they were reading. More importantly, many had learned to read with their eyes and not with their voice or fingers. Further, many of the participants started consciously to think about their reading proficiency and wanted to improve their rate beyond 300 words per minute. After this course in their second language, the students said that they were using such strategies in their first language also but were not aware of it. So, it could not be said that they were transferring such strategies from first language to second language.

Further, the strategies chosen in this study had implicit influences on improving the students' reading proficiency. Sight vocabulary has improved the students' ability of word recognition which helped them read fast. Metacognitive strategies have increased their own awareness of the reading process and improve the efficiency of these processes which are individualistic. These processes have also helped in the retention of what has been read and understood.

Strategies like using the contextual clues and guessing the meaning of words, predicting, re-reading have improved the students' ability to use their mental ability to interact with the text meaningfully. Such an interaction has enabled them to comprehend the matter quickly and retain it for a longer time. Ultimately reading is an interaction between the reader's mind and the text and if this activity has improved, the reader will be able to understand the text properly and through that the writer's idea would be revealed to the reader.

Major Findings

The reading speed of the students has increased steadily from one level to another. Though there has not been a regular uniform increase from one passage to another, there is a significant increase from passage 1 to 10. This seems to be of the strategies which they followed to read the passages. It can also be due to the attention given to them individually and the environment provided. But even after a gap of one month, the students had the interest to read and their improvement was evident in the last passage. They seem to have acquired the use of strategies while reading. When questioned after a period of time, they acknowledged the using of strategies while reading.

The comprehension level of the students has also improved from passage 1 to 10. This is evident from the increase in their comprehension scores from passage to passage. Complexity of the passage and the mental concentration of the reader seem to have a bearing on the comprehension level. But it is not certain if fast reading speed improves comprehension. Students who are moderate readers have comprehended better than the students who are slow or fast as revealed by the comprehension scores.

The retention level of the students has also improved. They were able to recall the content of the passages that they had read even after a gap of 15 days which was ascertained by an oral feedback. The students with higher level of involvement were able to retain more than the other students. The time devoted to think about the passages also seems to play a part in retention. When adequate time is spent on reflecting on the passage, retention seems to be better.

Conclusion

The study has concentrated on improving the reading proficiency of the students of engineering in the ESL classroom. About 36 students from rural background have been taken for the study and they have been instructed on the reading strategies to help them improve their reading skills. The reading speed, comprehension level, and retention capacity of the students have been analyzed and presented in this chapter. The students have shown good progress during this experimental study. They have got interested in reading when they have been made aware of its importance and how the use of strategies has helped them attain it. So even after the study, they have had an intention to read materials which interested them on their own. This interest and involvement have improved their reading skill. It can be assumed that this experimental study has exhibited that developing the reading proficiency of the students of engineering will help them acquire English language skills. It will help them to read and acquire the English language for their academic and non-academic communication needs.

Limitations and Suggestions for Future Research

The study was carried out only in one particular institution on an experimental basis. The study was carried out only among the students from regional language medium of instruction during their schooling. The participants chosen for the study were only boys in their first year of their B.E. program and were from villages and semi-urban towns. Further research can be carried out with the same set of students in their final year to find out whether they follow the strategies taught to them in the first year; to find out whether the strategy instruction has helped them to improve their comprehension in their technical subjects and if they apply these strategies in that domain. Moreover, the future research can also be carried out with only girl students or as a comparison between boys and girls, comparison between students belonging to rural students and students from urban towns or metropolitan cities, and with the students who have completed their schooling in English medium of instruction.

References

Aebersold, J. A., & Field, M. L. (1997). *From reader to reading teacher: Issues and strategies for second language classrooms.* New York: Cambridge University Press.

Ahuja, P. & Ahuja, G.C., (1991). *How to read effectively and efficiently.* New Delhi: Sterling Publishers Pvt. Ltd.

Anderson, J. Neil, Bachman, L., Perkins, K., & Cohen, A. (1991). An exploratory study into the construct validity of a reading comprehension test: Triangulation of data sources. *Language Testing, 8,* 41–66. doi: 10.1177/026553229100800104.

Barnett, M. (1989). *More than meets the eye: Foreign language reading: Theory and practice.* Englewood Cliffs, NJ: Prentice-Hall.

Bell, L. C., & Perfetti, C. A. (1994). Reading skill: Some adult comparisons. *Journal of Educational Psychology, 86*(2), 244–255. doi: 10.1037/002-0663.86.2.224.

Bond, G. L., & Tinker, M. A. (1957). *Reading difficulties: Their diagnosis and correction.* New York: Appleton Century Crofts.

Garner, R., Macready, G. B., & Wagoner, S. (1984). Readers' acquisition of the components of the text-look back strategy. *Journal of Educational Psychology, 76*(2), 300–309. doi: 10.1037/0022-0663.76.2.300.

Carrell, P. L., Devine, J., & Eskey, D. (Eds.). (1988). *Interactive approaches to second language reading.* New York: Cambridge University Press.

Chamot, A. U., & O'Mallety, J. M. (1994). Instructional approaches and teaching procedures. In K. S. Urbschat, & R. Pritchard (Eds.), *Kids come in all languages: Reading instruction for ESL students* (pp. 82–107). Newark, DE: International Reading Association.

Clarke, M. A. (1979). Reading in Spanish and English: Evidence from adult ESL students. *Language Learning, 29*(1), 121–150. doi: 10.1111/j.1467-1770.1979. tb01055.x.

Crandall, J., Jaramillo, A., Olsen, L., & Peyton, J. K. (2002). *Using cognitive strategies to develop English language and literacy. ERIC Digest.* Washington, DC: ERIC Clearinghouse on Language and Linguistics.

Dadour, E. S., & Robbins, J. (1996). University – level studies using strategy instruction to improve speaking ability in Egypt and Japan. In Rebecca L. Oxford (Ed.), *Language learning strategies around the world: Cross – cultural perspectives* (pp. 157–166). Honolulu: University of Hawaii.

Dubin, F., & Bycina, D. (1991). Academic reading and the ESL/EFL teacher. In M. Celce-Murcia (Ed.), *Teaching English as a second a foreign language* (pp. 195–215). New York: Newbury House.

Duke, N. (2003, March 7). Comprehension instruction for informational text. *Paper presented at the annual meeting of the Michigan reading association*, Grand Rapids, MI.

Emans, R., & Patyk, G. (1967). Why do high school students read? *Journal of Reading, 10*(5), 300–304.

Eskey, D. (1988). Holding in the bottom: An interactive approach to the language problems of second language problems of second language readers. In P. Carrell, J. Devine, & D. Eskey (Eds.), *Interactive approaches to second language reading* (Cambridge Applied Linguistics, pp. 223–238). Cambridge: Cambridge University Press. doi: 10.1017/CBO9781139524513.011.

Fuchs, L. S., Fuchs, D., Hosp, M. K. & Jenkins, J. R. (2001). Oral reading fluency as an indicator of reading competence: A theoretical, empirical, and historical Analysis. *Scientific studies of Reading, 5*(3), 239–256. doi: 10.1207/S1532799XSSR0503_3

Ghanta, S., (2019). Enhancing reading skills of engineering students. *International Journal of Recent Technology and Engineering (IJRTE), 8*(4), 520–521. doi: 10.35940/ijrte.D7218.118419

Globerson, T., Weinstein, E., & Sharabany, R. (1985). Testing out cognitive development from cognitive style: A training study. *Developmental Psychology, 21*(4), 682–691.

Goodman, K. (1970). Behind the eye: What happens in reading. In K. Goodman, & O. Wiles (Eds.), *Reading: Process and program*. Urbana, IL: National council of Teachers of English.

Goodman, K. (1976). Reading: A psycholinguistic guessing game. In H. Singer, & R. B. Ruddell (Eds.), *Theoretical models and processes of reading* (pp. 497–508). Newark, DE: International Reading Association.

Goodman, K.S. (1967). Reading: A psycholinguistic guessing game. *Journal of the Reading specialist, 6*, 126–135.

Gough, P. B. (1984). Word recognition. In Pearson (Ed.), *Handbook of reading research* (pp. 225–254). New York: Longman.

Grabe, W. (1991). Current developments in second language reading research. *TESOL Quarterly, 25*(3), 375–406. doi: 10.2307/3586977.

Gray, W. S. (1965). The nature and types of reading. In W. B. Barbe (Ed.), *Teaching reading: Selected materials* (p. 46). Oxford: Oxford University Press.

Hagley, E. (2017). Extensive graded reading with engineering students: Effects and outcomes. *Reading in a Foreign Language, 29*(2), 203–217.

Hathaway, G. M. (1929). Purposes for which people read: A technique for their discovery. *University of Pittsburgh School of Education Journal, 4*, 83–89.

Jensen, E. (1998). *Teaching with brain in mind*. Alexandria, VA: Association for Supervision and Curriculum Development.

Jensen, L. (1986). Advanced reading skills in a comprehensive course. In F. Dubin, D.E. Eskey, & W. Grabe (Eds.), *Teaching second language reading for academic purposes: Reading* (pp. 103–124). M.A. Addison: Wesley Publishing Company.

Kuhn, M. et al. (2006). Teaching children to become fluent and automatic readers. *Journal of Literacy Research, 38*, 357–388.

Kumari, M. J. (2016). Importance of reading and writing for engineering students in their technical journey. *AIJRELPLS, 1*(2), 72–76.

Leaver, B., & Oxford, R. (1996). A synthesis of strategy instruction for language learners. In R. Oxford (Ed.), *Language learning strategies around the world: Cross-cultural perspectives* (pp. 227–246). Manoa: University of Hawaii Press.

Myers, M., & Paris, S. G. (1978). Children's meta cognitive knowledge about reading. *Journal of Educational Psychology, 70*, 680–690.

Nuttall, C. (1982). *Teaching reading skills in a foreign language.* Oxford: Heinemann Educational Books.

O'Malley Michael, J., & Chamot, A. U. (1990). *Learning strategies in second language acquisition.* Cambridge: Cambridge University Press.

Oxford, R. (1990). *Language Learning Strategies: What Every Teacher Should Know.* Boston: Heinle and Heinle Publishers.

Oxford, R. (1996). *Language learning strategies around the world: Cross-cultural perspectives.* Manoa: University of Hawaii Press.

Paris, S., & Jacobs, J. (1984). The benefits of informed instruction for children's reading awareness and comprehension skills. *Child Development, 55*(6), 2083–2093. doi: 10.2307/1129781.

Sekar, Raja. (2006). *The effect of strategy instruction on teaching ESL reading.* Unpublished Ph.D dissertation. Hyderabad: CIEFL, Hyderabad.

Reading Today. (2008). *A Daily of International Reading Association.* Retrived from https://thereadingtub.org/reading-round-up-23-june-2008/

Rumelhart, D. E. (1977). Toward an interactive model of reading. In S. Dornic (Ed.), *Attention and performance* (pp. 573–603). New York: Academic Press.

Shearer, B. A., Ruddell, M. A., & Vogt, M. E. (2001). Successful middle school intervention: Negotiated strategies and individual choices. In. J. V. Hoffman, D. L. Schallert, C. M. Fair Banks, J. Worthy & B. Maloch (Eds.), *50th yearbook of the national reading conference* (pp. 558–571). Chicago: National Reading Conference.

Smith, F. (1982). *Understanding reading* (3rd ed.). New York: Holt, Rinehart and Winston.

Stanovich, K. (1990). Concepts of developmental theories of reading skill: Cognitive resources, automaticity and modularity. *Developmental Review, 10*, 72–100. doi: 10.1016/0273-2297(90)90005-O.

Stanovich, K. E. (1991). Word recognition: Changing perspectives. In R. Barr, M. L. Kamil, P. Mosenthal, & P. D. Pearson (Eds.), *Handbook of reading research* (Vol. 2) (pp. 418–452). New York: Longman.

Tsai, Y., & Chang, Y. (2014). Enhancing engineering students' reading comprehension of English for science and technology with the support of an online cumulative sentence analysis system. *SAGE Open, 4*(3), 1–9. doi: 10.1177%2F2158244014550610

Vogt, M., & Nagano, P. (2003). Turn it on with light bulb reading! Sound – Switching strategies for struggling readers. *The Reading Teacher, 57*, 214–221.

Widdowson, H. (1979). The process and purpose of reading. In H. Widdowson (Ed.), *Exploration in applied linguistics* (pp. 171–183). New York: Cambridge University Press.

Yoakam, G. A. (1951). The development of comprehension in the middle grades: Current problems of reading instruction. *Seventh annual conference of reading* (p. 32), Pittsburgh: University of Pittsburgh Press.

Part VI

Digital Education

Chapter 13

Techno-Mediated Teaching of English for Engineers in India

Perceived Threats and Hidden Opportunities

T. Ravichandran

With more than 10,000 engineering colleges in India including an intake of nearly 16 lakhs of undergraduate and two lakhs of postgraduate students per year, engineering curriculum, especially, the humanities, offer immense opportunities for teaching. Besides, with the young population attracted to novel technological gadgets, it should be much easier for the teachers to reach them by integrating their pedagogical methods through technology. To the contrary, many of the teachers are wary of using technical aids partially due to their perceived threats and mostly due to the inadequate support system available to train and empower them. Interestingly, those uncommon teachers who could use technology effectively rarely devote any time to share their experiences that will nudge the unwilling lot to embrace technical tools for effective teaching. This chapter, while filling this gap, by way of sharing the knowledge gained through first-hand experience, seeks to stimulate the apprehensive teachers to develop expertise in techno-mediated teaching. Simultaneously, it demonstrates how techno-mediated teaching facilitates reconfiguration of instructional design that focuses on the learner-centred mode whether it is for a limited 50 students in a cosy language lab or for 5000 and odd students managed through online course platforms like Moodle, Blackboard, or Brihaspati.

The Near Future Scenario: Inescapable Interpenetration of Technology

Girl: Dad, I'm in love with a boy who is far away from me. I am in Ghana and he lives in Tokyo. We met on a dating **website**, become friends on **Facebook**, had long chats on **WhatsApp**, he proposed to me on **Skype**, and now we've had 2 months of relationship through **Viber**. I need your blessings and good wishes daddy …

Dad: Wow! Really!! Then get married on **Twitter**, have fun on **Tango**. Buy your kids on **E-bay**, send them through **Gmail**. And if you are fed up with your husband … sell him on **Amazon**.

DOI: 10.4324/9781003268529-19

Jokes apart, the above conversation shared on a Facebook post prefigures the techno-dynamic time-flux that we are all caught in. While what has been once thought of as possible only in personal, intimate face-to-face contact such as dating, chatting, and thereby establishing human relationships, thanks to advancement in technology, which has actualised these relationships through virtual reality. Human touch, in a physical and figurative sense, has been glued to mobile touchpad screen or laptop keyboard. In this near future scenario where everything human can be transacted through online platforms, knowledge has become a malleable commodity. At this juncture, a teacher would be rendered obsolete if s/he refuses to embrace and adapt to the changes brought forth by technology. Adopting information and adapting to communication technological devices to reconfigure one's teaching methodology is the need of the hour. This has even more become imperative when the entire world has been physically locked down during the Covid 19 pandemic. Some of the teachers who refused to change their chalk-and-talk method lost their jobs, while those who quickly welcomed online meeting portals such as Google Meet and Zoom have enjoyed the challenges and made their teachings more effective than the offline mode. Besides, the paradoxical aspect of knowledge is that it is available for free, yet suitable guidance for using that knowledge is not free. Free information available on Wikipedia and YouTube is enough for many life-time learnings; yet, unless there is a teacher-facilitator to plan, guide, supervise, offer feedback, most of the times, the learner ends up with bewilderment and fatigue due to unbridled use of it. Excess of anything, as we all know, can be toxic to the mind and the body. Only the teacher can administer the right proportion like a doctor.

Is Technology a Friend or Foe of a Teacher?

Technology is anthemic to those teachers, especially from an Indian context, who still carry a Gurukul baggage and assume the unilateral power of a master disseminator who fills the empty minds of captive disciples with undisputed knowledge. While the Gurukul system had its significance when society believed in knowledge transfer and gain as an experiential lifestyle, and when technological tools were yet to usher in the traditional classrooms. Nonetheless, with the advent of multimedia technology, the old chalk-and-talk method of teaching supplemented by drills and homework done on notebooks is rendered obsolete. Computer technology enabled with course website that has a course calendar and sends automatic reminders through e-mails and messages, correlated with video lectures, multimedia tutorials, lecture transcripts, question banks, discussion forums, further links, and a compendium of soft copies of resource materials fosters alliance with both the teachers and the taught. In addition, when technological aids are used for quick assessment and feedback, it transforms the passive recipients in monocultural classroom climate to active participants. Thus, when the human teacher befriends the non-human technology in a seamless merging, deep learning with multiple modes and approaches is possible.

Disciple to Techno-Able

On the one hand, while teachers should adopt/adapt to and become adept in technological aids, on the other hand, they should also accept the paradigmatic shift in the mind-set of the students from a structured, disciplined, captive disciple mode to an amorphous, techno-abled, free, and boisterous style of fragmented receptivity. The present generation of students' attention span has been hijacked by social media and web-enabled e-weapons of mass distraction such as Facebook, Twitter, YouTube, and Instagram. In order to recapture the attention of the young minds and inculcate undivided devotion, the blackboard chalk-and-talk method is not adequate. Like knowledge that is compelled to mutate into digitalised forms for easily transmission and acceptability, the teachers ought to take their techno avatars to make their presence felt in the virtual reality where student community can be caught unawares in their meanderings. Unless and until teachers are willing to remould themselves and recast the fixed blackboard to dynamic mobile applications, they will not be able to encash on students' receptivity.

Techno-Push from Teacher-Centred to Student-Centred

While the permeability of technology compels a teacher to make technological aids indispensable in methodology, does it imply replacement of blackboard with virtual screen will promote deep learning? The question is rhetoric as those teachers who have just tried replacing conventional blackboard lecture on computer screen have already realised that they are ineffective. Merely dictating notes before the video camera, referring to online resource materials, making use of appealing visuals will not guarantee quality content delivered in an effective manner. The teachers ought to relinquish their all-knowing-ego and come down from their pedestal to reach out to the students by making the content tailor-made to suit the present needs of students. This implies changing the just-in-case rationale-based syllabi and curriculum to when-in-need model.

Ostrich-Like Wilful Ignorance is Not Affordable Anymore

Legend says that when the ostriches are scared, they tend to bury their heads on sand thinking that the troubles around will disappear on their own, and after some time, they can safely hold their heads high. The analogy is applicable to those teachers who foresee technological interventions as troubles in teaching methodology, and hope that they can resort to non-technological, conventional teaching methods soon. But they should realise that the times have irreversibly changed that they cannot afford to neglect technological advancements as temporary troubles. Nor is any redemption awaiting without technological aids. This is largely because of the path-breaking innovations and ramifications in cybertechnology, which have rendered human beings' cyborgs and have contributed to a sparklingly new cyberculture. The web, the internet, the mobile, iPad, YouTube, Facebook,

WhatsApp, Twitter, and other such new-age gadgets and applications have altered the way one thinks, communicates, writes, and responds. Keyboards, for instance, have drastically distorted the handwriting style of the humans. Similarly, the choice to surf through multiple channels on the television or browse through zillions of hyperlinks on the internet has considerably reduced the attention span of a normal human being. Hence, in this scenario, it is imperative for any teacher to comprehend the technologically diffused mind-set of the learners and teach accordingly.

Technological Appendages for Willing Teachers

The moment a teacher understands that technology is not a monster to be feared and abandoned but a toddler to be embraced and cherished, s/he is open to immense possibilities. Instead of looking at technological teaching aids as inscrutable demon-task masters, which would soon replace the human component, if the predisposed teachers perceive them as supportive tools for qualitative enhancement of teaching, they will be able to see them supplement as well as complement education. I would like to suggest few tools and software which I have been using as effective aids in my teaching. They are PowerPoint/Prezi, Microsoft Excel, Word, Google Doc, E-Mail (group/individual/course alias), Messenger (Yahoo, Facebook), Group chats (Facebook, WhatsApp), Piazza (for forums), and WordPress (blogs). They are less intimidating to start with and are user-friendly for both the teachers and the taught. All it needs a growth mind-set to try out new tools and a vision to expand teaching beyond the four walls of a classroom.

PowerPoint (PPT) and/or Prezi can be used for preparing slides that comprise not only words but also images, icons, pictures, and clip arts. A picture is indeed worth thousand words that adding illustrations makes a mundane blackboard writing more colourful and impactful. PPT is also a useful organisational tool for a teacher. The slides can be arranged or rearranged to suit one's thought-flow and plan of presentation. Besides, slides can be added every time a new idea crops up in the mind. Omissions and additions within a slide can be done in an effortless manner. Animations are used to make the items appear in a lively way. In pandemic times, where one is compelled to use recorded videos to students, PPT's easy-recording feature helps one to record and edit slide by slide. Excel sheet can be used for simple calculations to huge statistical analysis. Apart from supplementing Excel sheet for teaching, a teacher can efficiently use it for maintaining attendance, marks, and grades of the students. A Word document when it is shared in the form of Google document can be used for collaborative writing. Accordingly, project work assigned to a group of students can be executed easily on Google Doc.

Many teachers are hesitant to use their personal e-mail identities with students. Nonetheless, a teacher can always create a course e-mail ID that can be used for sending announcements, reminders, course policies, materials, and follow-ups. Besides, creating a course alias (ENG445), that is, a short name for the course group with all the registered students and tutors, helps in generating discussions and quick clarifications outside the classroom. Messenger services from Yahoo, Facebook, etc., are useful for instant communication and clarification. Instead of imposing the blackboard teaching method that is unable to sustain students' interests for a long duration, reaching out to students on the applications that they are familiar with makes them show involvement in classroom activities.

Group chats and discussions generated through Facebook, WhatsApp, as well as Piazza enhance the peer group learning process. In the class, the shy, introverted students, along with the backbenchers, never muster up courage to ask even genuine doubts in the class. Nevertheless, they feel empowered to ask questions in such group chats where one need not show one's face. Still, for a very reserved student, the group allows for anonymity. For an open and receptive teacher, the group chats have plenty to offer. The teacher here is not only a facilitator but also a learner. When the teacher is willing to create or join these virtual groups, students find it easy to relate to the teacher and absorb course content in a better manner. The burden of answering all the questions do not rest only with the teachers. Enthusiastic students answer many questions even before the teacher finds time to access them. In some forums, the teacher needs to mark only "the best answer", which encourages the students to give a careful thought before answering any questions. Owing to infinite possibilities of content sharing, the learners explore on their own, often, get an opportunity to learn from better material that even the teacher had never thought of.

Encouraging students to write blogs using free online sources such as WordPress fosters originality and creativity. Most of the times, when an assignment is given, students tend to copy, cut, and paste from online sources like Wikipedia. Making them write and submit assignments in the form of blogs compels a student to assess, assimilate one's own ideas. An interesting fact shared by a student who failed to submit course assignments in time is that blog-writing helped her to overcome writer's blog cum procrastination. She found that her mental inertia fizzled out the moment she realised that in a blog she can write anything instantaneously and spontaneously unlike the long and agonised wait for the perfect word to start with for the dreaded "assignment". Blog is also easy to assess and provide immediate feedback. The teacher, while letting each student write his/her blog without seeing others, can make all the completed blogs open after assessment.

In Line with Online Course Platforms

Teaching tools have proliferated than the appendages discussed above. While one can make use of the tools separately, one can save much time by resorting to online course platforms that have combined all these tools into one. At a basic level, online course platforms such as Sakai, Blackboard, Backpack, Google Classroom, Moodle, Brihaspati, and mooKIT are available freely. Most of them do not restrict the number of students so that one can effectively use them for managing Massive Online Open Courses (MOOC).

All these course platforms take teaching methodology to a higher and deeper level than what is possible in simple blackboard chalk-and-talk autocratic style of teaching. They allow separate login for teachers and users. Through the teacher's login, one can upload the course contents and teaching materials in the form of videos, audios, PPTs, PDFs, and Word documents. While these platforms appear to consume more time than conventional teaching at the initial stage, they are very helpful to the learners as they are student-centred. They can be used, as in the times of pandemic, entirely as a substitute for regular classroom teaching. Or they can be used for flipped method of teaching. Either way, learners benefit more as they get exposed to a multi-pronged learning experience.

In the ever-declining attention span, it would be difficult for a learner to pay full attention to long lectures stretched beyond 40 minutes. In such cases, learners make use of recorded lectures available on the online course platforms to watch the lectures at their own pace. One can always stop the moment one realises that the attention span is declining. Yet, at the same time, one can rewind any number of times to comprehend better, revise, and pause anywhere for reflection. Quiz questions that can be interpolated with the lecture videos help in checking through understanding of the concepts every now and then. The pace of the lecture can be manipulated by decreasing or increasing the speed of the videoplay.

In the conventional classroom, a new unknown word used by the teacher, unless written on the blackboard and emphasised, is likely to be missed by the students. This situation is averted as most of the videos are subtitled manually. Automatic subtitling is also available when they are uploaded in YouTube. Besides, subtitles, the transcripts of all the lecture videos are shared with the students. These transcripts help quick recollection before the preparation for exams. Transcripts clear misunderstood words owing to mispronunciation or idiosyncratic style of presentation.

Quiz conducted on these platforms though takes some time while creating the questions, save lots of time in evaluation as the system corrects automatically. Students will be happy to receive immediate feedback, while in the manual correction, the teacher not only takes time but also tends to commit mistakes in totalling. Assignments can be uploaded by typing in the given space or by uploading PDF and Word documents. An instructor can add tutors who can help in the evaluation of assignments. Marks will be displayed only to the concerned students thereby respecting privacy of individuals.

Overall macro-planning and micro-execution of tasks can be done effectively using these online platforms as they have inbuilt calendars and allows the instructor to arrange the materials, in a weekly, lecture-wise format. Many students feel that they have been kept in the dark in traditional classrooms where the syllabus gets unfolded as per the whims and fancies of the teacher. In the online platforms, the students can see the entire course content at one go and prepare themselves in advance. Apart from using all the above-mentioned online resources, the language laboratory can be used effectively to teaching communication skills, and especially, listening skill.

Using Language Laboratory for Teaching Listening Skills

CALL versus LALL

In the present age of information and cyber technology where communication is mediated through advanced tools facilitated by the computer, the Internet, and satellites, the use of language laboratory is considered obsolete and irrelevant. Computer-assisted language learning (CALL) has been replacing laboratory-assisted language learning (LALL) not only in highly developed countries like the USA and the UK, but also in some parts of India. While this change in the developed countries is effected through experiments, and theoretical shifts, in a developing country as India, CALL's popularity is merely on pragmatic considerations. It is easy to set up a CALL lab in India with as less as two lakh rupee investment, whereas, to establish a language lab would cost ten to twenty times more. The cost of the language lab set up depends on the number of booths. To have a single standard quality booth, imported from Sony, Japan, would cost around approximately one lakh. This may be apart from the cost involved in concealed cabling, false flooring, air-conditioning, and other accessories. Thus, even a language lab of 30 booths would cost a minimum of 30 lakhs. Of course, there are local products that cost cheaper than this but in the long run found to be defective and ineffective. Unlike this huge investment involved in establishing a language lab, CALL can be easily implemented in humanities departments by linking with computer centres. In this case, only software is to be procured for use in an established computer centre. Here, only time is to be allocated for language use in the computer centre. Even one high-end computer could serve the purpose when it is connected to any number of monitors.

The Need for LALL in India

Yet, there is a paradox prevailing regarding the use of language lab in our country. On the one hand, private institutions prefer CALL owing to its easy accessibility and less investment cost involved. In addition, the Indian universities, and technological institutes, in order to catch up with foreign trends, rationally opt for CALL. But, on the other hand, in the past decade or so, there is an increasing awareness at the state and central government levels in introducing language lab component as part of language teaching curriculum. The state governments of Maharashtra and Uttar Pradesh, for instance, have made the use of language lab mandatory in English language teaching curriculum. So, the paradox of the situation is that the function of language lab that is considered obsolete and luxurious is also thought to be relevant and useful. Today, especially when the country is experiencing an economic boom, for most of the upcoming institutes, the cost involved in the initial investment is immaterial as long as it serves a good purpose. Some educational institutes use language lab for ornamental purpose, that is, it is considered a show-piece and advertised for attracting students, yet most of the times, the lab remains closed. Of course, quite a few other institutes that genuinely see the need beyond the teaching of language skills but also for using the language lab for training students who are interested in competitive exams as TOEFL, SAT, GMAT, and others involving listening skills as a major component.

It is, in this paradoxical context, that I attempt to share some of my experiences in using a language lab for teaching English language and communication skills. Both as the Officer-in-Charge of the Language Lab at the Department of Humanities and Social Sciences, the Indian Institute of Technology Roorkee in the past, and as the Officer-in-Charge of the Language Lab at the Department of Humanities and Social Sciences, the Indian Institute of Technology Kanpur, and as an instructor of various courses related to language and communication, I would like to share my experience in the effective use of language lab for teaching listening as a significant component in communication.

Listening: An Ignored Aspect of Communication Skills

What is cumulatively thought of as "communication skills" is integrated with language skills that involve listening, speaking, reading, and writing. Nevertheless, compared to speaking, reading, and writing, listening as a skill to be trained is deplorably ignored. In fact, many language instructors nurture the notion among their students that communication skills can be developed without paying adequate attention to the listening aspect. Some do it owing to the fact that they do not find appropriate means of teaching listening skill. Even with language lab accessibility, they do not know that listening can be taught very effectively. To a large extent, many language teachers wonder whether listening skills can be taught at all. At this juncture, it is significant to note that language lab can be used for effective imparting of listening skill. It is equally important to know and accept the fact that listening is not only the basic but also the most important of all the skills needed for effective communication. A poor listener is bound to be a poor speaker, a casual reader, and ineffective writer.

The Need for Listening

It is the ability to communicate using language that essentially distinguishes human beings from animals. Man is gregarious by nature and even a prisoner who is sentenced to serve life term in solitary confinement longs for human communication. Effective communication is the key to success in personal as well as professional relationships. Many interviews were lost; many promotions were stalled; many relationships were broken due to miscommunication. However, communication does not merely mean "talking", it implies "listening" too. A successful communication process in which the message is carried from the sender to the receiver through a medium without any interruptions relies much on listening skills. Conversely, the ability to listen well determines the degree and intensity of communication. Unlike the common misconception, listening is not a *passive* quality. It is an *active* quality and integrated communication skill that demands energy, time, patience, and expertise. A research conducted on effective listening shows that communication involves 50% more listening than talking.

Listening Is Different from Hearing

The fact that listening is different from hearing is very nicely captured in a famous popular song by Simon and Garfunkel in the following lines of "Sounds of Silence":

> And in the naked light I saw
> Ten thousand people, maybe more
> People talking without speaking
> People hearing without listening ...

The singers actually satirise the modern society that has grown in terms of industrial developments and material goods but lost its human touch. People monotonously record faces and their voices but lose track of the messages.

Hearing is basically a *physical* activity whereas listening is predominantly a *mental* activity. During hearing, only the ears record sounds; but in listening process, the ears are the physical tools for recording sense, meaning, and perception. It is through listening, the mind starts *evaluating*, *recording*, and *interpreting*. One can *hear* sound even while sleeping. For instance, many music lovers are habituated to sleeping while leaving their audio systems on. However, one needs to keep the mind *alert* and *active* for listening. Effective listening demands complete involvement, full attention, and prompt response. The listener should give up any preoccupations, distractions, and day-dreaming.

Language Lab Facility

All language labs provide almost the same kind of facilities. The make by Sony, Japan, has been monopolising the market with its intricately insulated cables that prevents noise disturbance and maintains high-quality sound output. The model installed at the Indian Institute of Technology Kanpur is LLC5510MKII. It is a very old model yet fully functional. The number indicates the oldness of the version because now in the market LLC9000 series are available. Except for sophisticated soft buttons, minimisation of operations, and capacity for adding more booths the old lab equipment does not make any difference.

The Sony LLC-5510MKII is a language laboratory control console that permits the user to control up to 48 student recorders and the connected equipment for program sources. It has five function displays on its screen, which are easily produced by pressing the corresponding Function select switch on the right panel. By using these switches, the instructor can use various functions as master recorder selection, monitor speaker, audio selection, volume control, intercom, manual scan, all call, print out, etc.

The Three P's of Preparation

Using and mastering the console of a language lab requires lots of the three P's, that is, patience, practice, and perseverance. Those suffering from techno phobia (fear for technology) need not worry unduly, as the lab console is a very user-friendly one. Yet, before the use, the instructor needs to read the manual carefully. S/he needs to spend quite a lot of time in learning the operations of all the switches. Many language teachers shy away from the lab precisely for this reason, that is, it demands extra time. A one-hour session that involves intercom, pair, modelling, etc., would involve at least two hours prior practice by the instructor. However, the amount of extra time spent would be worthwhile once the instructor masters the technique. At the same time, insufficient practice can lead to improper handling of the console causing embarrassment before the student users.

Using Language Lab for Teaching Listening

Using language lab for teaching listening skills has distinct advantages. First, the closed environment facilitates thorough focus on the subject. Secondly, the headphone closely covering the ears safeguards the listeners from any unwarranted distractions. Thirdly, if task-oriented listening skills are set, then it makes the students fully concentrate on the topic. To set task-oriented learning modules, brilliant books and impactful audio-cassettes are available in the market. The one that I find quite effective is Marion Geddes' (1998) *How to Listen*. Although the book is subtitled as *An Intermediate Course in Listening Skills*, the book supplemented with an audio-cassette is an indispensable aid for teaching listening skills at any introductory level. I have used it to teach students at all levels of B. Tech. and M. Tech. students. Some lessons were used even while training teachers in refresher courses.

How to Listen

A recorded cassette that contains relevant programs from BBC accompanies this book by Marion Geddes (1998). The book contains the following nine chapters: "Why and how we listen", "How people speak", "Predicting what you'll hear", "Identifying important information", "How information is organized", "Repetition and rephrasing", "Making intelligent guesses", "Increasing your vocabulary", and "Following a news broadcast". A *Specimen listening comprehension test* follows these chapters. It also contains the *Tape transcripts* that help the instructor to check his listening skills before practising it for students. All the tests and exercises are also followed by an *Answer key* at the end.

The major advantage in using this book and the audio cassette is that it exposes the learners to English spoken by native speakers since the majority of the extracts were broadcast on BBC radio. Some students may initially find it difficult. The instructor is advised to play the cassette for more times until the students learn to catch up with the native speaking stress patterns and intonations. Nevertheless, the mere exposure with graded modules and self-study aids facilitates even weak students to enhance their listening skills by the end of the course. The specimen test devised for the Cambridge First Certificate Examination allows the students to gauge their expertise. It actually tests their listening skills in real-life situations as interview interactions, listening to announcements (particularly in airports), and locating a house using a map provided in a foreign location. Some of the lessons are quite motivating and interesting in terms of their contents that make the students look eagerly for more lab sessions.

International Business English

Another course that I could successfully try out in the language lab is *International Business English*. I have taught the course for final-year B. Tech. students as well as used some of its modules to teach Korean students in a short-term course. The course has been quite popular among the students for various reasons—it teaches certain other skills (telephone skills, personal interaction, writing reports, memos, meetings, etc.) required for communication at a professional level. Indispensable tools for this course are the books on *International Business English: Communication Skills in English for Business Purposes* prepared by Leo Jones and Richard Alexander (1989) and published by Cambridge University Press. The authors have prepared two separate books on the same topic to be used by the teacher (*The Teacher's Book*) as well as the students (*The Student's Book*). The books are accompanied by audio cassettes for use in the language lab.

Video Programs in the Language Laboratory

Those who have not made use of the language lab to the full extent think that the lab is meant for only audio cassettes. The lab can be used for video programs also. In fact, the booths are all equipped with monitors for video screening. There are so many video programs available in the market these days. Nonetheless, one video program that has been quite popular among IIT Kanpur students is *Bid for Power: English for Commerce and Industry* prepared by A. Fitzpatrick and C. St. J. Yates (1985). As the American, German, and Japanese business tycoons fight for a silicon project on the video, the students sharpen their ears to get the best out of the program.

Listening without Prejudice

I should complete this expose on language laboratory with a slight note of caution. Even when a receiver of a communication message has his ears tuned for receiving the message without any disturbance, there is no guarantee that the person listens to it without any prejudice. The following anecdote illustrates this very lucidly. Students at the University of California were asked by Mr. L. Agnew of the Department of Medical History for their reaction to the following:

> The father has syphilis, the mother tuberculosis. They have had four children—the first blind, the second died, the third was deaf and dumb, the fourth had tuberculosis. The mother is pregnant with her fifth child. The parents are willing to have an abortion. You have to make the decision.

That is, the students were asked to decide whether to have the fifth child aborted or not. Invariably, most of the students voted in favour of abortion. Mr. Agnew smiled and commented to them by saying, "Congratulations! You have just murdered Beethoven!" Ludwig van Beethoven was the deaf musical composer and child prodigy who gave his first public performance as a pianist when he was just eight years old! The anecdote shows how people are generally prejudiced listeners. Nevertheless, using language laboratory for teaching listening skills would always remain the first and foremost step in building up communication skills.

Concluding Thoughts to Ponder

The above discussions and demonstrations on the use of technology to enhance contemporary teaching and make it fully effective do not underestimate the significant role of a teacher—with or without technology. The teacher's primary role as a knowledge disseminator takes it for granted other soft skills related to leadership, communication, conflict resolution, empathic counselling, and motivating guidance. The aim of this chapter is to make it clear that information and communication aids facilitate the functioning of a teacher with all the implied skills in an effective manner. We all can easily recall that one teacher who taught much more effectively than without any technological aids—but the times were different and the students least experienced the onslaught of media distractions and ever-reducing attention span. While an excellent teacher will never lose her/his excellence for want of technological tools, outstanding delivery of content and deep impactful learning is possible if s/he is willing to complement her/his teaching with them.

References

Fitzpatrick, A., & Yates, C. St. J. (1985) *Bid for power: English for commerce and industry*. (Video Cassette with Teacher's Manual). London: BBC English by Television in Association with the English Language Teaching Development Unit.

Geddes, Marion. (1998). *How to listen*. Oxford: University Printing House. (with audio-cassette).

Jones, L., & Alexander, R. (1989). *International business English: Communication skills in English for business purposes—teacher's book & student's book*. Cambridge: Cambridge University Press.

Chapter 14

Integration of Zone of Proximal Development (ZPD) and ICTs in Language Learning

M. Ponmani and Geetha R

The zone of proximal development (ZPD) is the distance between the developmental level of learners' competence with guidance and without guidance. Vygotsky has strongly believed that the ZPD can transmit the learners from the social level to the individual level, i.e. from the inter-psychological level to intra-psychological level. Students of engineering are required to possess advanced technological skills required to compete in the digital and informative era. Though the students are able to readily handle the information and communication technology (ICT) tools such as mobiles, computers/laptops, etc., they are not able to differentiate and select the necessary, relevant, and authentic information effectively. This leads to their inefficacious performance in their learning process as well as at their workplace. In line with this, the chapter focuses on the relation between social and individual dissimilarity in current learning capability (an already achieved level) and probability (the achievable level with the help of others) of the students of engineering with specific reference to ICT. It is prognosticated that the future educational environment would be palpably digitalized besides the traditional teaching. This necessitates the teachers to enlighten the students with ICT tools for fostering the students' skill level and for capacitating them to excel at their workplace. The integrated ICT tools and ZPD promotes students' engagement in the learning process and effective comprehension. This integrated practice is predominant for the students of engineering as they deal with technical content and require adequate digital knowledge for successful career prospects. Eventually, the amalgamation of the ZPD and ICT in English classroom engenders the students of engineering to become self-directed learners (SDL).

Vygotsky (1978) expounded the zone of proximal development (ZPD) as, "the distance between the actual developmental level determined by independent problem solving and the level of potential development determined through problem solving under adult guidance or in collaboration with more capable peers". In other words, the ZPD is a state wherein the students are capacitated with the knowledge and skills that are requisite for autonomous learning (Figure 14.1).

DOI: 10.4324/9781003268529-20

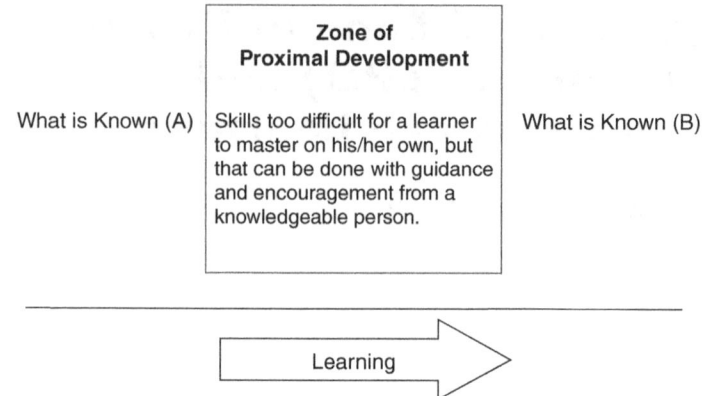

Figure 14.1 Zone of Proximal Development.
Source: Author.

The goal of the ZPD is to stretch the learners beyond their innate capabilities and challenging them, while providing support to meet their challenges. The ZPD is unique to each learner, and varies according to their capabilities, and the environment and the context wherein the development occurs. Apprenticeship and scaffolding are some common approaches to reach the ZPD (Collins, Brown & Newman, 1989). Scaffolding involves the construction of appropriate supportive conditions for a learner (Donato, 1994). In the case of scaffolding that is constructed for maintenance or repair of a building, it allows the workers to do jobs that they would not be capable of doing without climbing on the scaffolding. For example, scaffolding allows a worker to wash higher windows, provides a work surface for holding necessary construction tools, and often provides a level of safety and stability that gives comfort to the worker and thus the confidence to perform the job at hand. The term scaffolding may be used more abstractly in our context, but with familiar goals. Scaffolding for a learner consists of providing the appropriate tools, support, and confidence building, to allow the learner to succeed. When these types of models are employed, after spending time working with an expert to grasp a new concept, an individual is able to internalize what the learner has learned, and eventually scaffolding/assistance can be removed. This is reflected by Vygotsky, as he believed, "what the child is able to do in collaboration today he will be able to do independently tomorrow" (Vygotsky, 1989). In the ZPD, the expert engages in a "dialogue with the novice (learner) to focus on emerging skills and abilities" (Richard-Amato, 1988).

Another method in which knowledge can be passed from an expert to a novice is when the novice utilizes imitation to mirror what the expert has already mastered. However, this is only effective when the action being imitated is at the appropriate developmental level.

It is well established that the learner can imitate only what lies within the zone of his intellectual potential. The learner can enter into imitation through intellectual actions more or less far beyond what he is capable of in independent mental and purposeful actions or intellectual operations.

(Vygotsky, 1998)

As an example, describing how to blow up a balloon to a learner is a futile task. But while blowing up a balloon for the learner, the teacher should have noted how the learner began to purse his lips, puff out his cheeks, and blow out air through his mouth. Without an intended lesson, the learner has been imitating the actions of the 'balloon blowing expert.' Later, the teacher has witnessed the child blowing up the balloon independently.

The learner could not inflate the balloon as large as the adult, nor could he tie the balloon to keep it inflated, but incremental development has indeed been occurring to move him/her towards mastering these skills. This shows that these types of collaborations occur intentionally and unintentionally, they occur with teacher learner pairs, when mentoring and apprenticeship roles have been established, and also seemingly randomly through imitation or observation when the necessary context and environment are present. It has been noted that

the term 'collaboration' should not be understood as a joint, coordinated effort to move forward, where the more expert partner is always providing support at the moments where maturing functions are inadequate. Rather it appears that this term is being used to refer to any situation in which a learner is being offered some interaction with another person in relation to a problem to be solved.

(Chaiklin, 2003)

Teachers wish to utilize these ideas to make interaction with learners as effective as possible. According to Vygotsky, teaching "is effective only when it awakens and rouses to life those functions which are in a stage of maturing, which lie in the zone of proximal development" (Vygotsky, 1998). The expert is able to empower the learner through dialogue and guided participation (Rogoff, 1990).

English for Engineering Students

English is a tool that notably affects engineering students in academic life. While most of the assumptions in engineering are taught in English, it requires adequate proficiency in English language communication. In academic life, engineering students have to deal with the countless lectures, tutorials, labs, project reports, and papers in English. During the job-seeking process in interviews, GDs, it is mandatory to achieve mastery in English proficiency. After securing a job, they are entailed to work in groups since their task seldom be solved by an individual. A large number of Indian engineers have to now travel to many continents and work away from their domicile country. Also, among the scientists, technologists, and business experts from culturally and linguistically different communities, English has become the prime language for communication. So, being an engineer requires co-operating and communicating with different people from different parts of the world. English is used as the operational language on large extent. In order to harmonize with the colleagues, engineers have to speak fluent English. In this regard, English communication competence plays an important role in the academic life and career of engineering students.

Employability skills of fresh graduates have constantly received considerable attention in the local media. Lack of English language proficiency has often been cited as one of the major factors contributing to graduates' unemployment (Sharif, 2005). Jawhar (2002) has stated that in the private sector, graduates are becoming unemployable as a result of lack of proficiency in the English language. Various surveys have been carried out on employers in relevant industries to gauge whether graduates are meeting industry needs and the recurring theme that emerged from these surveys has been the lack of English language skills among fresh graduates and workers (MoHE, 2008; Tneh, 2008; Ambigaphaty & Aniswal, 2005; Sibat, 2005). These studies have implied that the majority of graduates and workers have been inadequate in English language proficiency, especially with regard to writing and speaking skills. This creates the need for the enhancement of English language proficiency of the engineering students. In line with this, the present chapter aims at addressing this need by integrating information and communication technology (ICT) and the ZPD in the engineering classroom.

Significance of ICT

Internet-based technology holds great potential for facilitating second-language teaching and learning, such potential can only be realized when this technology is actually used by the teachers and learners. The ways in which the English language learners (ESL) and teachers can make use of the technology will ultimately determine the effectiveness of this new approach in the teaching and learning of ESL. Therefore, a good understanding of the ways that teachers and the learners could make use of Internet-based resources is of great significance for both theoretical and practical purposes.

The term e-learning has originally described the use of ICT mainly in higher education settings, but it is now used to refer to Internet-based learning activities which have been wholly or partly incorporated into all levels of formal and non-formal education as course and curricular content (Tinio, 2003). The term 'blended learning' is used to describe a situation where ICT is used in conjunction with more traditional classroom-based activities. New generations of computer software, such as 'Blackboard' now facilitate online learning through interactive learning environments by combining learning and assessment materials and offering options for support and collaboration through communication with peers and teachers. Andrews and Haythornthwaite (2007) emphasize the possibilities for e-learning in the creation of communities of learning where knowledge can be generated, co-constructed, and presented as well as acquired. This involves engaging in a process of dialogue with other learners and teachers, where "learners and faculty communicate and work together to build and share knowledge" (Dwyer, Hiltz & Passerini, 2007). The idea of constructing and sharing knowledge accords with Vygotsky's (1978) constructivist propositions of learning.

ZPD in ICT Enabled ESL Classroom

The idea of ESL learners' imitating teachers can be a useful strategy for certain activities and learning tasks in student-centred language classes. Jones (2007) admits the value of 'teacher-led, repeat after me practice' in student-centred language classrooms and believes that this practice helps students to 'get their tongues around new phrases and expressions so that they can say them easily and comfortably'. Consequently, he recommends ESL teachers to employ this kind of student-centred practice.

Vygotsky's (1978) concept of the ZPD is also relevant to the application of Salmon's (2002) five-stage model of ICT learning. Salmon uses this idea in the 'scaffolded' (Bruner, 1985) approach to create tasks and activities which build on previous learning, and support the gradual extension of skills. Mercer (1994) judges that the relationship between teachers and learners is an important factor in determining the ZPD tasks, suggesting that the ZPD for a learner could change in different settings and with different teachers. Vygotsky's (1978) ideas on the role of teacher and imitation in the learning process imply a behaviouristic view of thinking. However, his ideas on learners' ability to self-construct knowledge, on the value of cooperation and interaction between teachers and students, and among students suggest a constructivist view (DeVries, 2000). It implies Vygotsky's belief in the possibility of thinking about the teacher-centred approach (behaviourist) and the learner-centred approach (constructivist) approaches as balancing one another, rather than being polarized. If teachers see this, they may integrate the strategies and practices of both these approaches in their classrooms.

Bruner (1985) and his associates coined the term 'scaffolding' as a metaphor for the kind of teaching support which builds on existing skills, guiding, and supporting the extension of these skills and is designed with opportunities for modification to suit learners progressing at differing rates and with differing support needs (Hammond & Gibbons, 2001; Salmon 2004; van Lier, 1996). Bruner (1996) also believed in the importance of the role of the teacher in the learning process, but stressed that the teacher should not monopolize the role, but should let learners 'scaffold' for each other as well. In the specific case of e-learning, this kind of support may be particularly significant; a number of writers have identified the potential for e-learners to feel isolated or invisible at times (McConnell, 2000; Pincas, 2004; Salmon, 2002). However, providing students with more scaffolding than they need may impact negatively on their learning. Teachers' understanding of the cognitive abilities of students can help them identify what tasks are suitable for them, how much support they need and when their scaffolding can be most effective. Fleming et al. (1998) suggested offering 'more structured support' for less capable students and more challenging tasks with less support for capable students in the language classroom. However, teachers' over-reliance on scaffolding may lead students to be more dependent, which can never lead to self-reliant and self-regulated learning. Freire (1984) discouraged the notion of what he calls 'assistancialism' in teaching and learning, because it would produce more adaptable learners and would enable teachers to practise more manipulation over students. However, teachers' careful selection of learning tasks in the light of their good understanding of students' cognitive abilities is an essential condition for enhancing students' involvement in problem-solving tasks and for encouraging their active learning. It is possible to use the Vygotskian 'Zone of Proximal Development' as a model for explaining the role of facilitator in a learner-centred classroom. For example, in a foreign language classroom, this idea can be used for distinguishing between the language tasks and challenging situations that students can perform independently from those that they can perform with assistance from teachers or more able peers. This accentuates the significant role of the teacher in the learning process and the importance of employing cooperative learning approaches in the classroom.

Cornelius-White and Harbaugh (2010) believed, "authentic, inquiry learning strategies balanced with direct instruction and cooperative learning are important methods in a learner-centred classroom". Kozma (2008) recommends changing the pedagogies appropriate to ICT learning in the following ways: "Pervasive technologies and social networks are used to support knowledge production and knowledge sharing by students and teachers. Networks are used to help students and teachers build knowledge communities" and "Teaching consists of challenging students to build on their knowledge and explore new topics. Collaborative projects and investigations involve searching for information, collecting and analysing data, generating knowledge products and communicating with outside professionals and audiences to share results". Thus, the idea of 'learner-centred' approaches to teaching underlines the collaboration and participation of learners through the whole process of designing course content and selecting learning procedures (Hedge, 2000). Learners are motivated to take responsibility for their learning. However, the teacher needs to ensure that effective strategies are chosen for learning and for monitoring the learning process. This approach is endorsed by Nunan & Lamb (2001) who assert,

"In an ideal learning-centred framework, not only will decisions about what to learn and how to learn be made with reference to the learners, but the learners themselves will be involved in the decision-making processes".

Gilly Salmon's Five-Stage Model

Gilly Salmon's five-stage model as shown in Figure 14.2 is chosen based on four reasons. First, this model was developed for e-learning and it supports social learning theory which capitalized on meaningful interaction based on Salmon's key premise. The premise holds that the learners' ability to learn online goes beyond the boundaries of technical aspects encompassing an underpinning social learning principle, where every individual surrounding the learner plays an important role in learning through his or her relationship with them under the support and guidance of a moderator. Second, Salmon's key premise on e-learning should be perceived as a result of interactions among learners, teachers, e-devices, knowledge, and the learning context (Abdullah & Saedah, 2010). Third, Salmon's model links

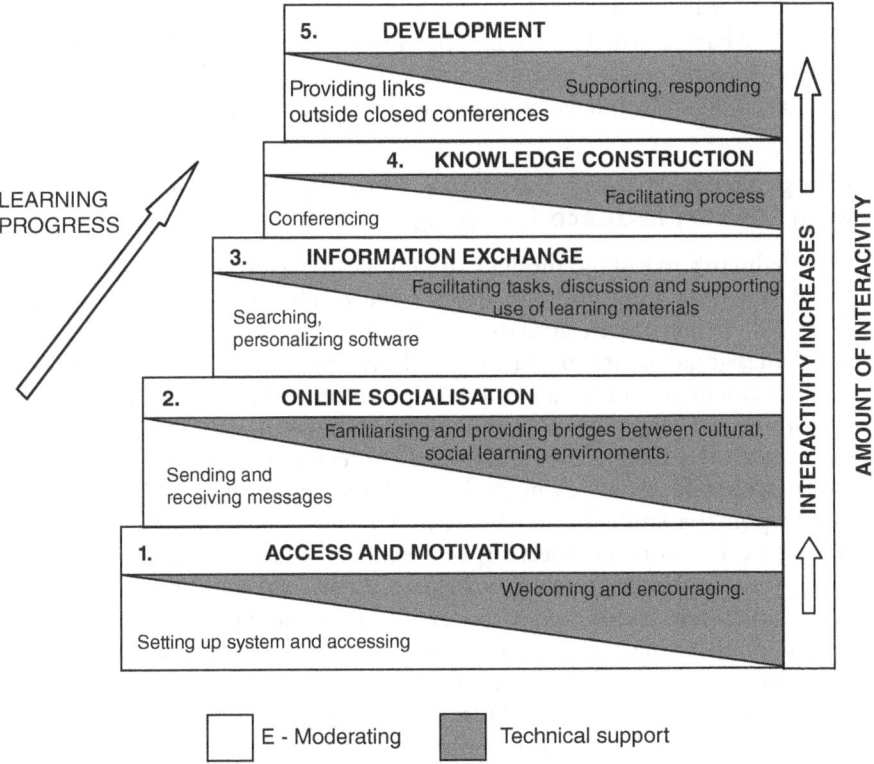

Figure 14.2 Gilly Salmon's Five-Stage Model.

(Figure Based on Salmon 2004).

to the theory discussed earlier as it could also be regarded as an extension of Vygotsky's ZPD (Attwell, 2006). According to the model structure, the moderator could gradually shift the responsibility of learners' development to the learning community guidance while the learner could eventually take charge on his learning by developing own scaffolding through relationship with other members of the community as well as beyond the community.

Another crucial reason why this model is chosen is that as e-learning should include informal learning, the key characteristic of the model shows how formal learning and informal learning can be interwoven in an e-context in aiding learners to reach their learning goals (Abdullah & Saedah, 2011). The Stage 1 of the model aims to encourage and guide students to participate in the online conference. Technical support is given to the students as the main focus task of the instructor. Stage 1 ends with students' first posting of message. Stage 2 aims to get students to establish their identities and initiate interaction and familiarize with it. Stage 3 is where interaction heightens with the use of learning management software for networking. In Stage 4, participants develop group discussions and collaboration among themselves in negotiation of knowledge and solutions to individual needs of learning. Finally, Stage 5 promotes individual reflection of what they have learned and achieved as well as critical thinking to advance to next learning goals. This e-learning-based model could settle in an undergraduate English language course of engineering colleges.

Integration of Gilly Salmon's Five-Stage Model with Undergraduate Language Learning

Gilly Salman's five-stage model can be used to investigate how this model could suit e-learning and can deduce the results by comparing activities done by students with activities in the model. This model can be used on 'Professional Communication Skills (PCS)' course, an undergraduate English language course; a compulsory subject to be taken in fulfilment of a four-year undergraduate study among engineering students. This course emphasizes the theory and practice of professional communication at the interpersonal level, in teams and to a large group. The course serves to build upon the students' academic and professional knowledge acquired through other core engineering or technical courses and aim to enable them to be highly effective in expressing themselves and in imparting their professional and technological expertise in a variety of jobs, business, and professional settings. The whole course was designed to last 14 weeks for each semester and divided into four parts: Process Description (Group Poster presentation), Technical Oral Presentation (Individual presentation), Business Meeting (Group presentation), and Persuasive Oral Presentation (Individual presentation).

In a conventional classroom, for the Persuasive Oral Presentation component, students will be given some lectures on guidelines on the effective persuasive speech aided by examples of authentic samples of effective and non-effective presentation. There are opportunities given to students to have a mock trial on persuasive presentation before their evaluation but due to time constraint and affective factors (embarrassment, lack of confidence, etc.), there is usually room for the two students to volunteer for the trial. Their presentation will be evaluated by the lecturer as a guide to other students but comments will be limited to students' individual presentation; in other words, the strength and weaknesses of a particular student may not be similar to other students who do not have the opportunity to present. All these will be on the first week of lecture for the component. The subsequent week until the end of the time will be allotted for this persuasive presentation component will be for evaluation of students' presentation. All students will be given one turn to present and receive their evaluation marks in the course. Students who get to present later will be more fortunate as they could learn from the strengths and weaknesses of their peers' earlier presentations as a guidance to present better. In other words, students will not largely be provided the opportunity to improve on their presentation output in the conventional classroom.

In the e-learning version of the course, students shall be briefed on the course design on the first day. There will be no physical classroom contact hours in the course (a great motivation for them to participate in e-learning) as they did not have to attend any classes. However, the students shall attend classes at the end of the course for summative evaluation by their lecturer. Learning would take place anytime or all the time wherever they are. Facilitator or moderator shall be the participant observer. Students may post and respond to messages, upload, and share notes and even videos of presentation. Students shall be encouraged to set up their own blogs and shall share it with the lecturer for observation purposes. These individual blogs, if set up would be for a particular scope of discussion shared through common interest among a particular group of students. Based on the five-stage model in Figure 14.2, tasks shall be given to the students as shown in the following Table 14.1.

This is to facilitate the observation in order to evaluate how the model fits in e-learning. The tasks shown are just samples to guide students and their lecturers in the conduct of their learning in e-learning environment. Students shall be evaluated for participation in the blogs (ongoing assessment) and on their final presentation (summative assessment) at the end of the course. Students' participations in blogs can be graded according to their level of engagement as described in Table 14.2. While analysing the content, facilitator or moderator shall concentrate on understanding the behaviour of learners which occurs naturally while interacting through the e-learning resources. E-learning resources promote natural interaction because students could assume anonymous identity and this allows them the liberty to express their ideas and feelings without social threats. Ideas and expressions would be authentic and this facilitates reliable data in gauging any aspects of students' learning skills. For example, a recent study conducted by Norlidah, Saedah, Mohd Khairul, and Hussin (2013) has revealed that the use of social blogs

Table 14.1 Sample Learning Tasks Based on Gilly Salmon's Five-Stage Model

Stage	Sample of Tasks
Stage 1: Access and motivation	• Accessing the blog by signing up and creating own username and password. • Lecturer shall initiate by posting the first message.
Stage 2: Online socialization	• Ensuring that the students have let others know their blog identity. • They may discuss about the power point slide on persuasive presentation and evaluation guide attached to the blog by posting questions or suggestions on the blog. • They may even discuss about their topic of presentation and elicit comments from other friends for suitability of the topic or improve on it.
Stage 3: Information exchange	• Study the evaluation form on what to expect from students in presenting an effective persuasive presentation. • Discuss and post suggestions on how to meet the criteria. • Focus on one criterion at a time. • Suggestions may in a form on a sample video presentation obtained from YouTube or even a website link to resources.
Stage 4: Knowledge construction	• By now students should have chosen a topic for their presentation. • They shall make a 7-minute video presentation, record it, and post it on the blog. • Elicit comments. • Facilitator or moderator may also comment other students' works to compare the presentations. • Identify common or shared problems or weaknesses which they want to rectify. • Students shall be encouraged to form a separate blog on their own with their friends and discuss the problem in detail.
Stage 5: Development	• The students may find the best solutions to their problems. • They shall be satisfied with your achievement. • They may share their experiences with their friends.

such as Facebook could enhance the creativity of Islamic Studies students in formal educational settings through their analysis on students, and their conversation exchanges through the social sites.

Implications of Gilly Salmon's Five-Stage Model and ZPD

Stage 1

This stage aims to promote individual learners' access and participation in social conference by welcoming them and providing technical support to

Table 14.2 Participation in e-Learning According to Level of Engagement

Level	Types of Engagement	Sample Activities
1	Observing and following	Observe and respond to learners either by asking questions or by adding comments.
2	Contributing	Actively upload and share information on effective presentation or even post own presentation to elicit comments.
3	Owning	Form or participate to discuss specific issues and problems shared commonly among them and seeking help either from peers or lecturers.
4	Leading	Take the initiative to seek other students having problems and offer assistance. This student could share the same problem and gather them to rectify the problem.

facilitate the use of e-devices in learning. Technical support could be shared with instructors, system providers, as well as from other learners.

Stage 2

This stage involves the students to establish their network identities rather than online identities and choose their social groups to interact by signing up as members or participants of a virtual community or even collaboratively creating own social groups relating to mutual learning needs with other individuals. Learning needs could be broad themes like socializing and writing or on specific needs like effective technical oral presentation or writing engineering research report.

1) Online is usually associated to connectivity to computer technology and communications specifically referring to states or conditions of device or equipment or other electronic functional units (Standard 1037C, 1996). This implies that online socialization relies on connectivity of computers and devices which is an essential factor for e-learning.
2) Network refers to a system of connectivity not only of technology devices, systems, and applications but also of people. In the adoption of Salmon's model, network implies a shift of focus on technology to the act of individuals forming and generating communication networks among themselves to interact, mediated by technology devices and applications. E-learning should place a primary focus on the learners; hence learner-centered learning. In the learning process, the learning environment and interaction among the learners should be the main foreground of the learners, whereas the technology devices should be the background.

Stage 3

Stage 3 involves initial scaffoldings to facilitate students' development in their presentation skills where they begin to interact and cooperate with each other by exchanging learning experiences to support individual's learning goals. The focus of learning here should be on creating sustainable networks of human interaction to facilitate learning needs where the technology devices qualify only as medium. Information exchanges could be initiated back and forth among social software.

Stage 4

This is the stage where interactions among students becomes more collaborative and students act on information shared in Stage 3 to form specific group discussions on respective subjects. Students embark on knowledge constructions in e-learning, they would also generate learning context on site that would also lead to more knowledge construction; the students may share common learning environment or context and develop the digital representation of the site or context. The site or context may not necessarily be a physical environment where the students is placed but could also be a network space or even a conceptual or abstract place such as a mutual learning subject or a learning problem (Nonaka, 1966).

Stage 5

This final stage is where the students reflect on what they have learned or acquired to help them achieve their learning goals. The reflection would lead to students' critical thinking to develop better or newer skills in developing higher competencies. For example, by reflecting on their learning process in Stage 4, learners would be able to understand better the elements to become better speakers and ways to utilize the new acquired skills to achieve their goals. This would also lead to new learning goals to develop further from their new acquired competency level. Another notable observation of students at this stage was that students became more responsible in their learning as they took charge in their own learning as their non-related aspect postings decrease significantly beginning at Stage 3 onwards compared to Stages 1 and 2.

Conclusion

Tools are the strategies used by a teacher in a classroom, so the learners can understand the concept more clearly. Vygotsky has provided us with a tool, 'the Zone of Proximal Development' that allows us to understand and enable learning. The ZPD is an excellent tool facilitating approximate and appropriate assistance to students to achieve a task. The ZPD promotes the teachers, parents, and mentors attached to a learner to recognize the learner's individual learning style. Further, the ZPD enables educators and parents to define the learner's immediate needs and the shifting developmental status, which allows for what has already been achieved developmentally, and for what the learner will be able to master in the future. Thus, it is evident that the integration of the ZPD and ICT would help the learners and teachers to achieve the goals and objectives of language learning in the engineering colleges. A good command over English language is all-time high demand skills in the present scenario of the 21st century. It is also considered a global language and inevitable for the development in this cutthroat competitions of the economies and professional world of developing countries. It is the only means to communicate and transact worldwide. The role of technology in making English language teaching and learning possible is the most crucial one. Using digital resources in schools, colleges, at home, and in societies, in general, has resulted in positive impact of expediting the teaching learning process and thereby being very effective and powerful in causing paradigm shift in pedagogy. Digital resources not just make teaching and learning interesting and easy but also help in developing critical thinking, inquiry-based learning, and teamwork in the case of both teachers and students while engaging in teaching and learning tasks. The dream of Global India can only be fulfilled by making immense and effective use of technology so as to empower English language of its citizens especially young learners. In concordance with these inferences, the present chapter has discussed the impact of Gilly Salmon's five-stage model to augment the engineering students' language proficiency. The chapter has recommended this model with the aim to integrate ICT and the ZPD in the English classroom of engineering students to enhance their English language competence and in turn, instigate students' self-directed learning (SDL) in the long run. This theoretical perspective will be proved effective when practically implemented in the engineering classrooms that the future studies could aim to conduct experimental study in this regard.

References

Abdullah, M. R. T. L., & Saedah, S. (2010). M-learning for future curriculum: Prospect and implementation. *International Journal of Multidisciplinary Thought*. Saedah, *1*(2), 1–11. ISSN 2156-6992.

Abdullah, M. R. T. L., & Saedah, S. (2011). The four C's of mobile capability as guiding principle for mlearning design: A shift of learners' focus away from technology. Malaysia: Masalah Pendidikan *(Issues in Education)*, Special edition. (pp. 105–114). ISSN 0126-5024.

Ambigaphaty, P., & Aniswal, A. G. (2005). *University curriculum: An evaluation on preparing graduates for employment*. Universiti Sains, Malaysia: National education Research Institute (IPPTN).

Andrews, R., & Haythornthwaite, C. (Eds.). (2007). *The Sage handbook of e-learning research*. London: SAGE.

Attwell, D. (2006). *Rewriting modernity: Studies in black South African literary history*. Ohio: Ohio University Press.

Bruner, J. (1985). Vygotsky: A historical and conceptual perspective. In James V. Wertsch (Ed.), *Culture, communication and cognition: Vygotskian perspectives*. Cambridge: Cambridge University Press.

Bruner, J. (1996). *The culture of education*. Cambridge, MA: Harvard University Press.

Chaiklin, S. (2003). The zone of proximal development in Vygotsky's analysis of learning and instruction. In A. Kozulin, B. Gindis, V. Ageyev, & S. Miller (Eds.), *Vygotsky's educational theory and practice in cultural context*. Cambridge: Cambridge University Press.

Collins, A., Brown, J. S., & Newman, S. E. (1989). Cognitive apprenticeship: Teaching the crafts of reading, writing, and mathematics. In L. B. Resnick (Ed.), *Knowing, learning, and instruction: Essays in honor of Robert Glaser* (pp. 453–494). Mahwah, NJ: Lawrence Erlbaum Associates, Incorp.

Cornelius-White, J. H. D., & Harbaugh, A. P. (2010). *Learner-centered instruction: Building relationships for student success*. Thousand Oaks, CA: SAGE.

DeVries, S. H. (2000). Bipolar cells use kainate and AMPA receptors to filter visual information into separate channels. *Neuron, 28*(3), 847–856. doi: 10.1016/s0896-6273(00)00158-6

Donato, R. (1994). Collective scaffolding in second language learning. In J. P. Lantolf, & G. Appel (Eds.), *Vygotskian approaches to second language research* (pp. 33–56). New Jersey: Ablex.

Dwyer, C., Hiltz, S., & Passerini, K. (2007). Trust and privacy concern within social networking sites: A comparison of Facebook and My Space. *AMCIS 2007 proceedings*, Keystone, Colorado, USA.

Fleming, P. J. S., et al. (1998). The performance of wild-canid traps in Australia: Efficiency, selectivity and trap-related injuries. *Wildlife Research, 25*(3), 327–338. doi: 10.1071/WR95066

Freire, P. (1984). Education, liberation and the church. *Religious Education, 79*(4), 524–545. doi: 10.1080/0034408400790405.

Hammond, J., & Gibbons, P. (2001). What is scaffolding? In J. Hammond (Ed.), *Scaffolding: Teaching and learning in language and literacy education* (pp. 1–14). Sydney: Primary English Teaching Assoc.

Hedge, T. (2000). *Teaching and learning in the language classroom*, 106. Oxford, UK: Oxford University Press.

Jawhar, M. (2002). *Education for the K-Economy: Challenges and Response* [Online]. Retrieved from http://www.sedar.org.my/articlePrint.cfm?id=16

Jones, K. S. (2007). Automatic summarising: The state of the art. *Information Processing & Management, 43*(6), 1449–1481. doi: 10.1016/j.ipm.2007.03.009

Kozma, R. B. (2008). Comparative analysis of policies for ICT in education. In *International handbook of information technology in primary and secondary education* (pp. 1083–1096). Boston, MA: Springer.

McConnell, D. (2000). *Implementing computer supported cooperative learning*. London: Kogan Page.

Mercer, C. W. (1994). Operating system support for multimedia applications. In *Proceedings of the second ACM international conference on multimedia* (pp. 492–493), San Francisco, California, USA.

Ministry of Higher Education, Malaysia (MoHE). (2008). The English language proficiency of Malaysian public university students. In Z. Mohd Don et al. (Eds.), *Enhancing the quality of higher education through research: Shaping future policy*. Kuala Lumpur: Putrajaya – The Ministry of Higher Education.

Nonaka, I., & Takeuchi, H. (1966). *The knowledge-creating company*. Oxford: Oxford University Press.

Norlidah, A., Saedah, S., Mohd Khairul, A. M. D., & Hussin, Z. (2013). Effectiveness of facebook based learning to enhance creativity among Islamic studies students by employing Isman instructional design model. *The Turkish Online Journal of Educational Technology, 12*(1), 60–67.

Nunan, D., & Lamb, C. (2001). Managing the learning process. In D. Hall & A. Hewings (Eds.), *Innovation in english language teaching* (pp. 27–45). London: Routledge.

Pincas, A. (2004). Practical principles of e-literacy for the e-society. *e-Society, 2004, 921*.

Richard-Amato, P. A. (1988). *Making it happen: Interaction in the second language classroom, from theory to practice*. NY: Longman Inc.

Rogoff, B. (1990). *Apprenticeship in thinking: Cognitive development in social context*. Oxford: Oxford university press.

Salmon, G. (2002). Mirror, mirror, on my screen – Exploring online reflections. *British Journal of Educational Technology, 33*(4), 379–391.

Salmon, G. (2004). *E-actividades: El factor clave para una formación en línea activa*. Barcelona: Editorial UOC.

Sharif, R. (2005). PC skills, English crucial, Kong tells grads. *The Star* OnLine. [Online] Available: http://www.icdl.com.my/news.asp?news=4 (03/02/2009).

Sibat, M. P. (2005). Leaping out of the unemployment line. Insite@unimas. *Teaching & Learning Bulletin, 6*, 2005. Retrieved from http://www.calm.unimas.my/insite6/ (12/11/2008).

Standard, F. (1996). 1037C. Telecommunications: Glossary of Telecommunication Terms. National Communication System. Technology and Standards Division. Washington, DC: General Services Administration.*Information Technology Service*.

Tneh, D. (2008, 15 September). Education, employment and the economy. *The Sun*. Retrieved from http://www.sun2surf.com/article.cfm.

Tinio, V. L. (2003). ICT in Education. Malina: E-ASEAN Task Force.

Van Lier, L. (1996). *Interaction in the language curriculum: Awareness, autonomy and authenticity*. London: Longman.

Vygotsky, L. S. (1978). *Mind in society: The development of higher psychological processes*. Cambridge, MA: Harvard University Press.

Vygotsky, L. S. (1989). Concrete human psychology. *Soviet psychology, 27*(2), 53–77.

Vygotsky, L. S. (1998). The problem of age. In R.W. Rieber Eed.), *The collected works of L. S. Vygotsky: Child psychology* (pp. 187–205). New York: Plenum.

Part VII

Teacher Education

Chapter 15

Towards Designing a Module for INSET

K. Manjula Bashini

The art of teaching is an entirely different skill, which requires an array of skill sets and a sense of commitment in the pedagogical process. So, teaching as a profession requires systematic training. As most teachers have no requisite methodical training of how to teach English in engineering colleges, they proceed with their own models. This chapter deals with the need for training teachers of English in the relevant methodology, which would optimally improve the learning of English by the students to meet their future academic and professional requirements. Though it is possible to hope for an English-speaking environment in a non-English-speaking country, many teachers of English fail to give their students a reasonable exposure to spoken English. The chapter not just establishes the need for In-Service Education and Training (INSET) but also sets up the priorities in designing a training module that is to be implemented in in-service training programmes for teachers of English. Moreover, this is an investment for the future well-being of the students and teachers as this deserves a high priority in the nation's education policy.

The National Education Policy 2020 strongly recommends Higher Education Institutes (HEIs) to provide larger multidisciplinary experience to its learners. Consequently, the empowerment of teachers' teaching competence becomes imperative to capacitate the learners towards the requisites of this knowledge age. In addition, studies have shown In-Service Education and Training (INSET) programmes enable teachers to focus on the need to enhance their academic competence and hone their professional skills for a meaningful classroom interaction with their students. The ability to appreciate learner needs and develop a stimulating learning environment in the classroom will enable teachers to contribute to a positive change in the teaching-learning process [NCATE (2010)]. As subject competency and strategies for imparting knowledge to learners are the two essential components of teacher preparation, effective teachers should visualize the learning context and apply appropriate strategies to help students learn better. Wingard (1983) remarked: "The need is most strongly felt when you have to teach is very different from what you have taught. This is the case when

DOI: 10.4324/9781003268529-22

a subject is changing rapidly or where the situation in which a subject is taught is changing rapidly" (244). Rubin (1978) observed that:

> The conception of new theory, the initiation of experimental research, the development of new materials and the improvement of instructional methodologies are useless, if their benefits are not incorporated into classroom procedures. Consequently, it is appropriate to review, from time to time, ongoing events to search for clues that will enhance the professional expertise of the practitioners.
>
> (ix)

It is relevant to review this growing body of literature on INSET and 'state of the art' pertaining to INSET, to help the readers to develop a perspective, which will facilitate the in-service training of the teachers of English in Tamil Nadu engineering colleges. Moreover, this review would facilitate to determine the priorities, and design a module for the in-service training and education (INSET) of teachers of English working in engineering institutions in Tamil Nadu.

Genesis of INSET in India

Chaudhary (2002) believed that ESP could have appeared as early as the eighteenth century with the purpose for trade and administration in British India (39). Moreover, it was in 1844 Lord Hardinge, the then Governor General, declared that Indians who knew English would get preference in employment [Chaudhary (2009: 477)]. Correspondingly, English for employment was one of the prime objectives of teaching English in colonial India. But soon after independence, regional languages were made the medium of instruction in schools and colleges. This led to the deteriorating standards in the teaching of English and caused a great concern to the central government. These were the reasons for appointing several 'committees and study groups' from time to time, and to review the teaching of English in India and make recommendations. As a result, appropriate teacher training and re-training schemes at tertiary level of education were introduced. The Kothari Commission (1960) of India observed that:

> For a successful completion of the first-degree course, a student should possess an adequate command over English, be able to express himself in it with reasonable felicity, understand lectures in it and avail himself of its literature. Therefore adequate emphasis will have to be laid on its study as a language right from the school stage. English should be the most useful 'library language' in higher education and our most significant window on the world.
>
> (15)

Based on the review of the Study Groups (1967 and 1971), the Government of India made many recommendations for improving the quality of teaching English at all levels of education. The American Commission on Teacher Education rightly observed,

> The quality of a nation depends upon the quality of its citizens. The quality of its citizens depends not exclusively, but in critical measure upon the quality of their education, the quality of their education depends more than upon any single factor, upon the quality of their teacher.
>
> (Mohanty (2007: 273))

The need for training collegiate teachers of English has been taken up by the Central Institute of English and Foreign Languages (CIEFL), Hyderabad, set up by the Central Government for bringing about qualitative improvement in the teaching at the tertiary level. Many teachers in Tamil Nadu have benefitted from attending the in-service courses organized by the CIEFL. But it is impossible for one institute to provide for the INSET needs of teachers of English of an entire nation.

Moreover, the provisions made by the UGC for the in-service training were applicable to teachers of English who taught in arts and science colleges. The number of engineering colleges was very few that they did not draw any attention. But today, the extensive growth of engineering colleges has outnumbered that of the arts and science colleges and the worth of engineering education in the prevailing job market cannot be underestimated. Despite the significant role that English plays in our educational system, the aims and objectives of teaching English in engineering colleges are far from being ideal. In recent years, the universities in Tamil Nadu have introduced changes in the syllabus seeking to make the study of English more relevant and useful for students. Some of these changes relate to the communicative approach in language teaching. The teachers of English in Tamil Nadu engineering colleges need to be given an orientation regarding the changes in the teaching methodologies introduced to improve the proficiency of students to meet the globalized needs. Therefore, it is necessary for state governments (here the Tamil Nadu government) to take the initiative to design and execute appropriate training programmes.

So, these observations, particularly relevant in the context of teaching English in Tamil Nadu engineering colleges today, with no requisite specific training expected from the teachers' side, to address a larger mass of mixed ability engineering group of students, emphasize that there is a need for the study. A very large majority of teachers working in engineering colleges have had no training in English teaching methodology. Besides, these teachers are brought up on a mainly literature-based syllabus and a lecturing mode of instruction. So, they do not seem sensitive to the changed circumstances, in which vast numbers of students lack the minimum English language abilities required to follow lessons in English. The dire necessity in the new context is an approach to teach the language, adopting techniques of English which in turn would facilitate the learning of English congenial at the tertiary level. Further, Gokak's (1961) statement, "There is a crisis in the teaching of English in India…Teachers recruited in recent years are not themselves always sure of their command of the English language" (1) is still relevant.

A Profile of Collegiate Education in Tamil Nadu

A brief description of the collegiate education in Tamil Nadu will help in understanding some problems to consider the priorities of INSET for teachers of English in engineering colleges. There are about 552 engineering colleges (government, government-aided, and self-financing) in Tamil Nadu as on July 2014. There is a combined strength of more than a thousand teachers teaching English mainly to first-year BE/B Tech students, and in many colleges, they are teaching English to third-year students also. The data was collected from the teachers in the cadres of Assistant Professor, Associate Professor, Professor and Head of the Department. Earlier, the minimum qualification for appointment as a college teacher of English was only a first-class or high second-class MA degree in English. For many years now it has become mandatory for aspiring teachers to possess a pass in National Eligibility Test (NET)/State Level Eligibility Test (SLET) or a degree in PhD. Now AICTE has made these as necessary qualifications to apply for a teaching assignment. There is no teaching qualification, like B Ed or Post Graduate Diploma in Teaching English from EFLU, prescribed for those wishing to join college teaching, or any requirement once having joined, to acquire a teaching qualification. The insistence on M Phil or PhD, together with improved opportunities for teachers to work for these qualifications, has resulted in most teachers obtaining higher academic qualifications, but no preparation in teaching methodology. It is relevant to mention here that, rather strangely, there is no emphasis on candidates with professional qualifications like PGCTE/PGDTE offered by the internationally acclaimed English and Foreign Languages University (EFLU), Hyderabad.

A number of in-service training courses for the teachers of English in Tamil Nadu colleges conducted in the last few decades were mostly for the permanent teachers in arts and science colleges, usually conducted by the academic staff college affiliated to the concerned universities. Likewise, permanent faculty of premier institutes like National Institute of Technology (NIT) attend orientation programmes at National Institute of Technical Teachers Training Institute (NITTR) during their induction to service. These were ad hoc programmes which rarely provided teachers with prior information regarding the content and methodology of training adopted for the courses. It must, however, be admitted that the teacher-trainers sometimes on their own initiative tried to ascertain the participants' views concerning the problems of teaching/learning at the tertiary level. This was generally done in large group discussions in which, of the 40 to 50 teachers present, only a small number took an active part, the rest being contented to remain passive listeners [Sivaraman (1984)]. It is noted from the interviews conducted with experts that no systematic survey of teachers' perception of their INSET needs was ever undertaken in the past, to initiate more positive steps in the training of teachers based on their perceived needs. This study, therefore, took up empirical survey of teachers' views and their reactions to specific aspects of INSET in the context of Teaching English as Second Language (TESL).

Taking into account, the role of English in higher education, there is a dire necessity to raise and maintain its standards in the engineering colleges and universities. A number of teacher orientation courses for teachers of English in higher institutes of learning sponsored by the University Grants Commission (UGC), All India Council for Technical Education (AICTE) of India, and the Tamil Nadu Government were conducted for the benefit of tertiary-level teachers of English. Among these important programmes were the UGC Summer Institutes in English, and Faculty Improvement programmes of UGC and also of the Tamil Nadu State Government, and the short in-service courses financed by the State Government. These programmes touched only the fringes of the problem of training large number of teachers. However, except for the solitary UGC Summer Institute in English for Specific Purposes (ESP) (1986) in M.S. University Vadodara, Gujarat, no major initiative exclusively intended for teachers of English in engineering colleges was undertaken anywhere in India. It must be mentioned, however, the National Institute of Technical Teachers' Training and Research (NITTR) Chennai conducted short courses of five days duration for the benefit of teachers working in engineering and polytechnic colleges. EFLU, the country's only university for English studies and teaching has also to the best knowledge of this researcher not taken in providing opportunities for teachers of English in engineering colleges. Therefore, this study focused on setting the priorities in the in-service training of teachers of English in engineering colleges in Tamil Nadu. A large section of these teachers never had the benefit of specific training for teaching English in engineering colleges.

The review of literature related to INSET, in general, and developments in Teaching English as a Foreign Language (TEFL)/Teaching English as a Second Language (TESL) in particular helped to develop an appropriate INSET for teachers of English in Tamil Nadu engineering colleges. An empirical survey of views of teachers and students on teaching/learning of English in engineering colleges in Tamil Nadu was undertaken for this study. The data was collected through a questionnaire survey from teachers and students in engineering colleges. The responses received from teachers and students were statistically analysed to obtain insights relating to the study of English in these colleges, and, specifically, to the teachers' perception of their INSET needs. In this regard, teachers' views on the teaching of English in engineering colleges reinstated their training needs and the significance of the organization of specific in-service training courses. Interviews were conducted with teacher trainers and their views/suggestions were duly considered in drafting the training module for the teachers of English in engineering colleges.

Methodology

A pilot survey was conducted by administering 20 questionnaires to teachers of English working in engineering colleges in and around Tiruchirappalli, Tamil Nadu. Based on the outcome of the pilot study, a new teacher questionnaire was constructed and the content validated by the senior ELT advisers and experts, Dr. K. N. Sivaraman (Former Director, ELT, Government Higher Education Department, Tamil Nadu) and Dr. C. Meenakshi Sundaram, Former Professor, Department of Humanities and Social Sciences, National Institute of Technology, Tiruchirappalli. Care was also taken to ensure that the questions were unambiguous and required clear-cut answers. At the same time, clarity in instructions was ensured, to assist individuals respond appropriately to the questions. The data was collected through questionnaire from selected colleges of engineering and technology in Tamil Nadu through cluster sampling technique. The questionnaire survey was administered through mail, and personal interaction with teachers. Moser and Kalton (1971) remarked, "Mail surveys with a response as low as 10% are not unknown… and strenuous efforts would be needed to bring the response rate above 30 or 40%" (262). In the present survey, of the 275 questionnaires sent to teachers of English in Tamil Nadu engineering colleges, 153 were returned to the researcher. Considering the difficulties involved in the mail survey, this response rate of 55.63% is very satisfactory. Likewise, 700 student questionnaires were distributed and 578 were received, out of that 532 completed questionnaires were taken for statistical analysis. The responses of questionnaires were coded for statistical analysis. An interview guide for the purpose of interviewing the English Language Teaching (ELT)/English for Science and Technology (EST) experts was developed. The interview guide was also subjected to content validation by two experts Dr. K. N. Sivaraman and Dr. C. Meenakshi Sundaram. The schedule consisted of ten queries. The discussions in the interview helped the researcher in reassuring the need and usefulness of in-service programmes for teachers of English and also in suggesting a draft module of the syllabus for the in-service programmes for teachers of English in engineering colleges. The data collected for the present study were processed and analysed with the help of Statistical Package for Social Science (SPSS) Version 17.0. The researcher has used descriptive analysis, Pearson's correlation coefficient, one-way ANOVA, and t-test for further analysis.

Findings of the Study

Some of the findings of the study are:

- It is observed that both the teachers and the students consider the students' objectives of learning communication skills in English is for being job ready and for maintaining good interpersonal relations, and also for their future professional and academic requirements. So, it could be inferred that both teachers and students believe, learning of English will nurture students' language for library use, preparation of project reports, study skills, soft skills, etc., which in turn will result in their job readiness. Hence, it is clear that teachers and students have understood the significance of effective communication skills and have common objectives when it comes to the students' learning of English language.
- Based on the prioritization by teachers, the survey reveals that a large proportion of teachers believe that their own lack of awareness in recent ELT/ESP methodologies and lack of teaching competence stand as vital factors in affecting students' learning of English in engineering colleges. It is evident from their ranking that many teachers have realized that they are lacking specific training and aptitude towards teaching, which are very important in the context of teaching English in engineering colleges. Accordingly, the self-realization from the teachers for the want of training programmes is a welcoming change and reflects their positive approach towards learning.
- Teachers consider themselves fairly proficient in their productive skills, when compared to their receptive skills. There can be no better outputs of speaking and writing, without the language inputs in the form of reading and listening. Hence, it is a saddening fact that teachers indirectly acknowledge that they do not possess good reading and listening skills which are crucial requisites for ELT practitioners.
- Though teachers are confident on possessing some of these language skills, they do not express the same level of confidence when it comes to teaching the same skills to their students. This reinstates the fact that teachers need to be fully equipped in their own language skills and the methodologies to teach such skills to their students.
- Teachers are not aware of the relative merits and deficiencies in the various teaching methodologies. It is obvious that they are not updated with the recent developments in ELT in general, or in India, in particular. It is evident that teachers do not possess the essential knowledge of ESP/EST methods and approaches required to teach English for students of engineering and technology.
- Almost 75% of teachers have affirmed that it is essential for teachers of English in engineering colleges to undergo in-service training. Most of the teachers strongly believe that in-service training is crucial for their career and it will help them become proficient teachers of English in the long run.

- Teachers' preference to topics like teaching project preparation, technical writing, group discussion techniques, resume writing, etc., to be included in in-service course modules reveals that they feel that training in the above-mentioned topics would enable them to equip themselves with the requirements and meet the present-day demands of their students.

One of the findings of the survey concerns the strong emphasis on an appropriate integration of theory and practice in the content of the in-service courses. Strevens (1977) said, "the purpose of the theory element in teacher-training is to provide 'understanding' as distinct from 'knowledge'" (79). Teachers are in favour of large group discussions followed by lectures by fulltime teachers. Although teachers rated supervised teaching practice by trainees as an important activity, in general, they appear to be reluctant to participate in small group discussions and practical work in small groups. Thus, the study throwing light on the teachers' low rating of the importance of small group discussions in in-service courses may be due to their lack of self-confidence in speaking in English in a group of teachers and teacher-trainers. This attitude of the teachers could be changed with their attending of INSET programmes. The findings of the study strongly reveal the need for INSET of teachers of English in Tamil Nadu engineering colleges.

Views of the Teacher Trainers on the Need for Training Teachers of English

This study was strengthened with valuable insights from a series of interactions/interviews with a senior ELT adviser and expert, Dr. K. N. Sivaraman, an Emeritus Professor, a national resource person for UGC Summer Institutes in ELT and ESP, with vast experience in designing and conducting INSET programmes for college teachers of English in Tamil Nadu. The first-hand information obtained from this expert about the UGC Summer Institute in ESP for teachers of English from engineering and polytechnic colleges held in Maharaja Saproji Rao University, Vadodara, Gujarat, in 1986, the only one of its kind, has been invaluable for this study. Further, this expert also provided the useful information regarding an initiative through the distance education mode for training teachers of English in colleges of engineering and technology to work for a diploma in teaching English for Specific Purposes. This initiative, unfortunately, was given up for lack of administrative and academic support, and also a lack of enthusiasm on the part of teachers themselves. Further, in order to know the views of other renowned teacher trainers in the field, an interview was conducted and following are the views shared by them:

- Teachers of English do not possess the required level of proficiency and competency in both using and teaching the language.
- The current English syllabus in engineering colleges do not reflect the objectives of teaching English in engineering colleges.

- The materials used for teaching English in engineering colleges are quite inadequate to achieve the aims and objectives of teaching/learning English.
- In-service courses designed to improve teachers' language skills and teaching competency would certainly help them to become better teachers.
- Hence, universities and colleges must make it mandatory for teachers to enhance their teaching competence through on-going training programmes which have adequate theory and practice sessions.

Some of the suggestions made by the experts in the field are:

- 'Lack of interest in developing a professional attitude to teaching ESP' Dr. K. N. Sivaraman (Former Director, ELT, Government Higher education Department, Tamil Nadu).
- 'The methods of teaching English in engineering colleges do not contribute to realising the objectives' Dr. T. Sriraman, Retired Professor, EFL University, Hyderabad.
- 'Teachers from rural colleges need intensive training' Dr. V. Janakiraman, Director, Nethaji Subash Chandra Bose College for women, Tamil Nadu.
- 'In-service training needs to be goal-specific and voluntary' Dr. Paul Gunashekar, Professor, EFL University, Hyderabad.
- 'In-service training could include case studies to be shared with engineering students' Dr. Julu Sen, Professor & Head, Department of ELT, School of Distance Education, EFL University, Hyderabad.
- 'In-service training courses should be meaningful and intensive. And it should be made obligatory' Dr. S. Mohanraj, Dean, English Language Education, EFL University, Hyderabad.
- 'Teachers should hold accountable for the desired outcome of students in the teaching and learning process' Dr. C. Meenakshi Sundaram, Former Professor, Department of Humanities and Social Sciences, National Institute of Technology, Tiruchirappalli.

Guidelines for Implementation

Private universities in Tamil Nadu, like Vellore Institute of Technology, Vellore (VIT), Tamil Nadu have established separate training wings called Academic Staff College (ASC), to provide ongoing training programmes for their faculty. But the focus of the training programme is on general communication skills, pedagogy on teaching and learning, knowledge of ICT and other engineering subjects, for all teachers across the various disciplines. There is no programme offered specifically for the teachers of English. Similarly, Krishna Group of Institutions (colleges of engineering and technology), Coimbatore, Tamil Nadu, have planned a separate division called

Continuous Professional Development Centre (CPDC). In similar to VIT, the management has massive plans of organizing training programmes for all the teachers of engineering and technology throughout Tamil Nadu. They have planned to run courses for engineering teachers similar to that of B Ed and M Ed courses for school teachers. Their proposal too does not envisage any special programmes for teachers of English in engineering colleges. The Teaching-Learning Centre (TLC) National Institute of Technology, Tiruchirappalli, initiated by Technical Education Quality Improvement Programme (TEQIP II), was established with the main objective of enhancing teachers' teaching and communication skills, using ICT tools. The TLC at NIT-T have organized invited lectures, short-term courses, workshops, and faculty development programmes across disciplines. But it is limited to institute faculty members that could be developed as full-fledged resource centre which can extend its training wing to nearby engineering colleges. The scope of TLC could be expanded to the language domain and skill-based training could be provided to excel the teachers in the art of teaching.

According to the recommendation of the Second Study Group (1971: 15–16) "there should be an English Language Teaching Advisory Committee to advise on how teacher training resources should best be deployed and applied, and to decide what levels … what areas need attention" [cited in Sivaraman (1984)]. Based on this, a committee of ELT specialists could be formed including professors from EFLU and experienced professors from other colleges to facilitate the work of TLCs in co-ordination with the nodal centres to organize and evaluate the training courses. The funding for all these proposed INSET courses could be from the AICTE and the MHRD (India). Already AICTE is sponsoring programmes on 'Pedagogy and Communication Skills', a programme for teachers of all disciplines. This could be extended for training the trainers of teachers of English so that they could disseminate the art of edification across disciplines. The implementation of such programmes is complete only with proper periodical evaluation and assessment.

Priorities in Designing a Module for INSET

The INSET programme should provide adequate scope for teachers to improve their theoretical knowledge as well as practical skills required for an effective teacher of English for Specific Purposes. A significant problem in the design of teacher-training courses in the past, according to Marsh (1981) "has appeared to be a matter of identifying an appropriate mix of theory and practice in a range of prescriptive packages…" (4) [Cited in Sivaraman (1984)].

INSET programmes should,

- assist teachers in improving their command of language, wherever necessary, and in developing fluency and accuracy, in the use of 'English for Teaching Purposes';

- assist teachers in developing an awareness on the developments of ELT/ESP methodology;
- enable teachers to acquire a variety of techniques and classroom strategies for ESL teaching in engineering colleges, and provide ample opportunities for experimenting with the methodology in appropriate contexts or situations;
- encourage teachers to examine their work and provide opportunities for discussing their own and their colleagues' approach to ESL teaching in engineering colleges;
- identify teachers with potential for further intensive training which will enable them to become teacher-trainers.

[Willis (1981a: 198) cited in Sivaraman (1984)]

Likewise, while planning the course, priority should be given to the 'time' factor. Gower (1983) observed that: "... There is so much you want trainees to be able to do and so little time for them to learn to do it in, that if you are not careful, your targets equally become unrealistic" (210). So, it is important for the in-service course to "have limited aims, a clear direction, and form a connected whole", (op cit) which would enable the teachers to develop a greater awareness of the theory and practice of ELT appropriate to their work place requirement. With this view, the INSET for teachers of English in colleges of engineering and technology should provide opportunities for them to attend, mandatory short intensive courses of two or three weeks duration which can be organized in suitable nodal centres and strategically located in Tamil Nadu for about 30 or 40 teachers in each centre. Provision of such courses will lead to an increase in the number of teachers trained in ELT methodology. In addition, the provision of a home-study module for teachers selected to undergo in-service training could also be offered. The advantage of home-study module is that teacher-trainers will be able to give the participants a clear idea of what to expect in the in-service courses. Teachers, then, will have a prior view of the theoretical preparation and practical work they will be exposed to in the two or three weeks intensive course. The home-study module will include basic study materials for learning, allowing enough time for teachers to complete a few assignments and return them to their tutors for comments. It would be useful to include a short reading list, to sustain the motivation level of the teacher participants.

Proposal for Implementation

A sample syllabus of the training module with guidelines for implementation of INSET for teachers of English in colleges of engineering and technology in Tamil Nadu is proposed here. Prior to suggesting a sample syllabus, it is necessary to make some general observations with regard to syllabus design. There are some general factors which apply to teacher-training syllabuses:

- A syllabus must help the teachers to achieve certain specific goals.
- A syllabus must have a starting point (where the teachers are in terms of their knowledge, ability, and experience) as well as an end point, so it implies movement from one position to another.
- "A syllabus is a practical instrument, and it is judged by its success or failure". So, it must fit in with the realities of the in-service situation.
- A syllabus is an organization of the material taught and activities involved in the course. So that, if required, the general principles and the specific organization can be discussed and improved.
- In order to make teacher INSET effective, it will be necessary to make the above principles explicit.
  [Gower (1983: 146–147) cited in Sivaraman (1984)]

In concord with these statements, Brumfit (1984) expounded: "The basic principles, underlying assumptions, and objectives of any training programme should be made as explicit as possible, because if they are not ... they will prove impossible to object to, adjust and improve" (114).

Training Module

It is proposed to give a brief description of a module which may be covered in the two weeks of the INSET. Continuous follow-up sessions or refresher courses will have to be arranged every five years, in addition to the orientation courses for teachers of English joining engineering colleges subsequent to the launching of the INSET programmes, within two-week time (a ten-day programme planned excluding Saturday and Sunday) of four sessions a day (three theory and one workshop session) accounting to 40 sessions. The syllabus, broadly speaking, will comprise the following sections.

INSET module can be categorized into the following major areas:

A. Language Teaching Methodology
B. Language for Technical Communication
C. Curriculum Design and Materials Development
D. Strengthening Classroom Practice

Section A: Language Teaching Methodology

This section will include the study of:

1. Approaches and Methods in Language teaching
2. Communicative Language Teaching

3. English for Specific Purposes
4. Testing and Evaluation

1. Approaches and Methods in Language Teaching

   Introduction to Theories of Language teaching/learning – Audio-Lingual Method – Grammar Translation Method – Total Immersion Method – Task Based Language Teaching – Content Based Language Teaching – Communicative Language Teaching – Computer-Aided Language Teaching

2. Communicative Language teaching

   Language and Communication – Importance of four language skills – Knowledge of Language and its use – Communicative competence – Development of Skills and Strategies – Strategic competence – Need for development of Strategies

3. English for Specific Purposes

   English for Science and Technology- English for Specific Purposes – English for Academic Purposes – English for Occupational Purposes – Development of ESP – ESP Learners – Needs Analysis – Course Design – Role of ESP Teacher – Use of Language in Different Contexts – Register and Jargon

4. Testing and Evaluation

   Understand the key concepts and the basic principles of language testing and assessment – Testing as a teaching procedure – Role of testing and evaluation in L2 curriculum – Construct and design a test for a specific skill – Evaluate various assessment approaches appropriate for language learning – Current trends and problems in language testing.

Section B: Language for Technical Communication

This section will include the study of:

1. Teaching of specific language skills
2. Technical communication
3. Aspects of technical writing

1. Teaching of specific language skills

   Teaching of Listening Skills – Teaching of Speaking Skills – Teaching of Reading Skills – Teaching of Writing Skills – Teaching of Study Skills – Teaching of Dictionary and Reference Skills

2. Technical Communication

Difference between General and Technical Communication – Characteristics of Technical Communication – EST discourse – rhetorical functions – techniques – technical exposition – technical narration – description and argumentation – Preparing for Placement Interviews – Presentation Skills – Group Discussion Techniques – Negotiation Skills – Problem Solving Skills – Interpersonal Skills – Soft Skills and Life Skills.

3.   Aspects of Technical Writing

Characteristics of technical writing – Fundamentals of grammar – language and style – rhetorical features – advanced writing – writing letters and memos – report writing – feasibility reports – progress reports – evaluation reports – lab reports – drafting project proposals.

Section C: Curriculum Design and Materials Development

This section will include the study of:
1.   Teacher as researcher/collaborator

Expected learning outcome – Analyzing target audience needs – Design a course/syllabus – Designing specific modules – Methodology to be adopted – Activities relating to target discipline – Assigning and implementing project-based learning/case studies – Project works – Role of ESP practitioner

2.   Teacher as material developer

Materials production – Designing authentic materials – Designing supplementary materials for classroom interaction – Designing tasks – Selection of material – Learner-centered materials – Teacher as decision maker.

3.   Teacher as evaluator

Evaluation procedures – Summative and formative assessment – Criteria of a good test – Test specification – Types of tests – Placement test – Achievement test – Diagnostic test – Proficiency test – Grading and scoring – Alternative assessment – Portfolios – Journals – Self/Peer assessment.

Section D: Strengthening Classroom Practice

This section will include the study of:

1.   Teacher Orientation

Methodology and strategies of teaching – Current trends and directions – Classroom interaction – Motivating and managing learners – Managing learner difficulties

2. Training of ESP teacher

    Academic Integrity – Educational Psychology – Continuing Professional Education – Enhancing Professional Practice – Micro Teaching – Practice Teaching – Supervised Field Experience – Demo Sessions by Experienced Teachers

3. Panel Discussion and Workshop

    The discussions with renowned English Language Teaching experts and teacher trainers in the field have enabled the researcher to bring out the sample module for INSET courses. However, in drafting the syllabus, the researcher has also referred to Landon (1988), Mekala (2004), Sivaraman (1984), and TESL (2016).

Conclusion

The findings of this survey led to the determining of priorities in TESL INSET and to the development of the proposal for the INSET. There is a need for specific research in different aspects of TESL training in Tamil Nadu. This would lead to qualitative improvement in the study of English at the tertiary level. Trained teachers of English in engineering colleges can motivate the students by illustrating with case studies on how good communication skills help engineers to improve their career prospects. They can help students to identify their own shortcomings in the four language skills. This awareness can be customized and strategized according to their regional environment to design innovative tasks for their students. Hence, a trained teacher is equipped with an exercising skill of lifelong learning and form an impetus factor to impart outcome-based learning modules. Therefore, INSET programmes will empower teachers to impart the employability skills in their learners, as the role of English in the socio-economic development of India is undeniable [Mishra (2013)].

References

Brumfit, C.J.. (1984). *Communicative methodology in language teaching: The roles of fluency and accuracy*. Cambridge: Cambridge University Press.

Chaudhary, Shreesh. (2002). The sociolinguistic context of English language teaching in India. In Shirin Kudchedkar (Ed.), *Readings in English language teaching in India* (pp. 37–66). Chennai: Orient Longman.

Chaudhary, Shreesh. (2009). *Foreigners and foreign languages in India: A sociolinguistic history*. New Delhi: Foundation Books.

Gokak, V. K. (1961) Mid-service training for teachers of English. *Bulletin of the Central Institute of English, Hyderabad, 1*, 7–15.

Gower, R. (1983). Saving time in teacher-training. In R. R. Jordan (Ed.), *Case studies in ELT* (pp. 210–220). London: William Collins Sons and Co. Ltd..

Landon, John. (March 1988). Teacher Education and Professional Development. *Tesl Canada Journal/Revue Tesl du Canada, 5*(2), 56–69.

Marsh, G. (1981). Introduction. In British Council (Ed), ELT Docs – 110 focus on the teacher (pp. 4–6). London: British Council.

Mekala, S. (2004). *An alternative syllabus for students majoring in english literature: An ESP framework*. Germany: Lambert Academic Publishing Limited, 2011. ISBN 978-3-8454-1364-8.

Mishra, Uday Kumar (2013) Learning English for development and economy in India: A socio-cultural perspective. In Dr Philip Powell-Davies, and Professor Paul Gunashekar (Eds.), *English language teacher education in a diverse environment*, Selected papers from the third international Teacher Educators conference Hyderabad: ELTAI. 16–18 March.

Mohanty, J. (2007). *Teacher education*. New Delhi: Deep and Deep publications, 273.

Moser, C. A. & Kalton, G. (1971). *Survey methods in social investigation*. London: Heinemann Educational Books, 262.

Rubin, L. (1978). *The in-service education of teachers: Trends, processes and prescriptions*. Massachusetts: Allyn & Bacon Inc.

Sivaraman, K. N. (1984) Priorities in the in-service training of teachers of English in Tamilnadu Colleges. University of Manchester. Manchester *Unpublished M.Ed Dissertation*.

Strevens, P. (1977). *New orientations in the teaching of English*. New York: Oxford University Press.

TESL. (Fall 2016). 401/601: English Language Teaching I (ELT I). Washington, DC: American University.

The Kothari Commission. (1960). *Report of the Education Commission Appointed by the Government of India under the Chairmanship of D.C. Kothari*. New Delhi: Government of India Publications Division.

The National Council for Accreditation of Teacher Education (NCATE). (2010). *What Makes a Teacher Effective?* Retrieved January 11, 2017, from http://www.ncate.org/Public/ResearchReports/TeacherPreparationResearch/WhatMakesaTeacherEffective/tabid/361/Default.aspx.

Willis, J. (1981a). The training of non-native speaker teachers of English: A new approach. In British Council (Ed), *ELT Documents 110 – Focus on the teacher* (pp. 41–53). London: British Council.

Wingard, P.G. (1983). Training English lecturers in an Overseas university. In R.R. Jordon (Ed.), *Case studies in ELT*. London: Collins ELT.

Chapter 16

Reflective Practice for Teachers in the Post-Method Era

S. Soundiraraj and B. Andria Babu

In the post-method era, teachers have the choice to select the methods that are suitable to their classroom situations and the requirements of students. No single method can be suggested as the ideal one for teachers because the key features of various methods may be collectively useful in dealing with typical classroom contexts and the needs of the students. Therefore, teachers have the autonomy to decide the appropriate teaching practices for their students, which indicates the key point that they should possess the academic expertise in the selection of methods, material preparation, and design of language activities. Reflective practice is a viable solution in the post-method era to enable teachers to be autonomous in language classroom. It can be described as way to evaluate and analyse their past actions and incidents to get insights for effective teaching learning situation. Gibbs model of reflective practice (1988) that has six steps like description, feeling, evaluation, analysis, conclusion, and action plan can be effectively used to enhance the professional skills of the teachers to meet the demands of the post-method era. This chapter describes how reflective practice using Gibbs model can be exploited to make the teachers autonomous in different aspects of second-language teaching.

Reflective teaching is a beneficial practice that makes teachers think about the methodology and approach of their teaching. Reflective practice has attained the major focus of interest among teachers in the post-method era. It is also a process where teachers reflect on their teaching practices, investigate how something has been taught, and how the style of teaching can be improved for achieving their goals. It contributes a lot in professional development for both pre-service and in-service teachers. Ur (1999) mentions that the most important basis for professional progress is based on teachers' own reflection on daily classroom events. She lays emphasis on professional development of teachers through reflective practice on language activities which they conduct in language classes.

Bartlett (1990) points out that becoming a reflective teacher involves moving beyond a primary concern with instructional techniques and asking questions like 'how to', 'what', and 'why'. These instructional and managerial techniques are not the ends in themselves but as part of broader

DOI: 10.4324/9781003268529-23

educational purposes. Asking questions such as 'what' and 'why' adds positive energy to the individual's teaching, resulting in the emergence of autonomy and responsibility in the work of teachers. The sincere efforts taken by teachers to find out the answers for the above said questions based on the analysis of their classroom teaching will make them more autonomous in the choice of methods, materials, and teaching techniques for better teaching learning process. The knowledge gained by teachers through reflective practice improves learner performance as well as teacher development.

The major reason for reflective practice is that teaching experience alone does not necessarily lead to effective learning but essentially requires deliberate reflection on experience. Experience without the application of the reflective practice will be unproductive as far as the effective teaching is concerned. It is clear that teaching experience needs to be linked with reflective practice for desirable results. Moreover, it is a significant approach in practice-based professional learning settings. Here teachers learn from their professional experiences, rather than from formal learning or knowledge transfer.

In the present context, it is essential to know how far the teachers of English are aware of the usefulness of reflective practice and find out whether they rely on reflective teaching to enhance the language learning process. The present study aims to give a clear picture of the reflective practice used by teachers in the current scenario. For this study, google form questionnaire has been employed, to collect necessary data from 50 teachers of English working in various engineering colleges of Tamil Nadu, India. In addition, it provides a brief account of different reflective practice models, which may serve as background knowledge needed for the study.

Models of Reflection

There are different models of reflection that attempt to describe various stages of reflective practice. A brief summary of some prominent models such as (a) Kolb reflective cycle, (b) Gibbs reflective cycle, (c) Schön model, (d) Driscoll model, and (e) Rolfe et al.'s reflective framework may give a comprehensive idea about reflective practice. This will be useful to select a model that is appropriate for the study.

Kolb Reflective Cycle

Kolb (1984) developed a model known as experiential learning cycle which consists of four stages like concrete experience, reflective observation, abstract conceptualization, and active experimentation. The first stage is about an individual's experience of any situation in everyday life. In the second stage, he reflects or reviews on the experience based on the particular situation. This will help him form an idea or concept, which is categorized as the third stage 'abstract conceptualization'. In other words, what he derives from reflecting on his experience is either an insight or conclusion. In the fourth stage of cycle, he uses the insight gained from his experience in different life situations. In this stage, he is an active experimenter to find out whether his new insights are effective in various contexts. The four-stage model is relevant in teaching context since it helps teachers in reflecting on and arriving at conclusions from a hands-on experience that they have while teaching language in classrooms. It also promotes planning to do something innovative and testing it out. It combines everyday experience with evidence-based educational research.

Gibbs Reflective Cycle

Gibbs reflective cycle developed by Graham Gibbs (1988) presents a structural representation for an individual's learning from his experience. This cycle has six stages such as description, feeling, evaluation, analysis, conclusion, and action plan. It explains the process of learning from experience focusing on six different stages. Each stage has its own significance in view of the process of learning from experience. The first stage is about the description of the experience and the second one is related to the feelings and thoughts that happen as a result of the experience. Then, in the third stage, the individual evaluates the experience objectively and looks at both the positive and negative aspects. The next stage is 'analysis', in which he attempts to get meaning or insight from that experience. In addition, this stage is concerned with reasons for the success or the failure of a particular situation. The fifth stage is known as conclusion, which is about the skills or knowledge one acquires from it and what one could have done differently for achieving the expected outcomes. The final stage of this reflective cycle is 'action plan' which he designs to deal with similar situations successfully in future. While discussing the importance of reflective practice, Gibbs (1988) says:

> It isn't essentially adequate to have an experience to learn. Without reflecting upon this experience it might rapidly be overlooked, or its learning likely lost. It is from the feelings and thoughts rising up out of this reflection that generalization can be created and it is this generalization that permit new circumstances to be handled successfully.

Schön Model

Schön (1983) formulated a model that explains two different types of reflection, viz. reflection-in-action and reflection-on-action. Reflection-in-action happens when the action or work is in progress. As a result, changes can be made to achieve the desirable outcomes. But, reflection-on-action involves reflecting on the past experience of a situation that an individual has faced recently. In other words, reflection-in-action occurs during the work, whereas reflection-on-action occurs after the completion of the work.

In the case of reflection-in-action, it is necessary to focus on the following aspects – (a) situation, (b) decision, and (c) immediate action. First, the situation has to be monitored and decision needs to be taken to improve the situation if there is a need. Finally, immediate action to bring about changes in the particular situation is carried out. These are the major features of reflection-in-action. On the contrary, reflection-on-action involves the following steps – (a) reconsidering the past situation and (b) identifying the necessary changes that can be brought about in similar situation for the future. Schön model comprising the concept of reflection-in-action and reflection-on-action is useful in language teaching contexts as it paves the way for effective teaching.

Driscoll Model

Driscoll model of reflection has been developed by Driscoll (2007) taking into account three major questions – 'what', 'so what', and 'now what' raised by Terry Boston in 1970. He has incorporated these three questions as three different stages of an experiential learning cycle. The first stage is meant for describing an event or incident and it is connected with the question 'what'. In fact, the description answers questions like (a) What happened? (b) What was your reaction? (c) What was the reaction of other people? The second stage is known for the analysis of the incident and it is related to the key question 'so what'. This stage has the space for answers for the questions like (a) How do you feel now about the incident/situation? (b) What is the impact of the incident on you? (c) What is the insight that you have gained from the incident? Finally, the third stage is meant for the action that has been proposed after the incident or experience. The proposed action is evolved as a result of questions such as 'What should happen to change the situation?', 'What is your plan to deal with similar situations for the future?', 'What will happen if you do not implement the changes?' The third stage of this experiential cycle is linked with the key question 'now what'. Thus the Driscoll model based on three questions raised by Terry Boston is suitable for teachers of English who strive for effective teaching.

Johns Model of Reflection

Johns Model of Reflection (1995) is a reflective practice model framed by Christopher Johns, which is based on the seminal work of Carper (1978). In this model, there are two major aspects, viz. 'internal focus' and 'external focus' in the process of thinking, which are given much importance. According to Johns, 'internal focus' denotes emotions and thoughts of a person, whereas 'external focus' indicates either situations or incidents. Moreover, there are five phases in this model, which is presented in the form of cyclic process. The process starts with description of the experience and then moves towards the second phase of reflection. In the third phase, the emphasis is on influencing factors like internal and external focus. The fourth phase is meant for the answer to the question 'Could I have dealt with it better?' and the final one in the cyclic process is known for the learning experience of the individual.

All the five models discussed here are more or less similar in the theory of reflection. Though there are different models of reflective practice, Gibbs model of reflective practice has been selected for the present study since it covers the prominent aspects of all the reflective models. In addition, the six stages mentioned in the Gibbs model are very clear and they can be easily incorporated in the questionnaire used as a tool in the study to collect information related to the reflective practice of the teacher participants.

Related Studies

A number of researchers have examined how reflective practice can be used for professional development and effective language teaching. Impedovo and Malik (2016) emphasize reflective practice and research attitude for the professional development of teachers. They argue that reflective practice is a continuous process of learning from experience; teachers by using reflective practice need to remain lifelong learners for effective teaching-learning environment. Sellars (2012) highlights the point that teachers are the prominent people who bring about educational changes despite the role of policy makers, curriculum developers, and education authorities. Besides, the quality of educational changes is based on the teachers' capacities for reflective practice and the development of self-knowledge. While discussing the usefulness of reflective practice, Smyth (1993) suggests that it should focus not only on technical skills but also on ethical, social, and political contexts in which teaching happens.

Al-Ahdal and Al-Awaid (2014) put forth the view that teachers who rely on reflective teaching and critical reflection, improve their quality of teaching by updating their awareness and making changes not only in their attitude but also in the structure and delivery of their lessons in the classroom. Robichaux and Guarino (2012) conclude that reflective habit plays a vital role in the professional growth of pre-service teachers. Their study confirms that professional growth of pre-service teachers who have been involved in portfolio assessment have been better than those who have not been involved in portfolio assessment. This is due to the use of portfolio assessment that inspired the participants to rely on reflective practice.

Priya Mathew et al. (2017) suggest that reflective practice with techniques like recording the class and maintaining diary/journal can be used effectively for the professional development of student teachers. The study by Xhaferi and Xhaferi (2012) reveals that reflective practice can be used as a tool to improve the learning and critical thinking skills of the students as students have positive attitude towards reflective journal writing. All these research studies discussed here highlight the point that reflective practice has positive effects on professional development, which is useful for maintaining effective teaching learning situation in language classes.

Methodology

For the present study, the researcher chose 50 teachers of English working in various engineering colleges of Tamil Nadu, India. Among the 50 teachers, 31 were male, whereas 19 were female. The responses regarding the reflective practice of these participants of the study were collected through google form questionnaire consisting of three subdivisions, viz. (a) personal profile, (b) views of teachers on reflective teaching, materials, and methods, and (c) reflective practice of the respondents. The data collected through the questionnaire as a tool was analysed using simple percentage analysis.

Hypotheses

The hypotheses framed for the study are as follows:

1) Teachers recognize the usefulness of reflective practice in enhancing their quality of language teaching.
2) Teacher autonomy in language classroom can be ensured when teachers adopt reflective practices for effective teaching.

Analysis of Data and Interpretation

Personal Profile

Among the three sections of the google form questionnaire, the first section is related to the personal profile of the teachers. As per the participants contacted for the study, 62% of them were male and 38% of them were female. In view of teaching experience, teachers were classified into four categories – (a) Category-1 with 1–5 years; (b) Category-2 with 6–10 years; (c) Category-3 with 11–20 years; and (d) Category-4 with more than 20 years. Table 16.1 displays the percentage details of the experience-wise classification of teachers. This shows that the third category has the highest percentage of the teachers with 11–20 years of experience, whereas the fourth category has the lowest percentage of teachers with more than 20 years of experience.

Table 16.1 Experience-Wise Classification of Teachers

Sl. No.	Experience	No. of Teachers	Percentage
1	1–5 years	11	22%
2	6–10 years	15	30%
3	11–20 years	20	40%
4	Above 20 years	4	8%
Total		50	100%

Views of Teacher Participants

The second section of the google form questionnaire has six questions to know the views of teacher participants on reflective practice, methods of teaching, and materials. The analysis of their views will be helpful to know how far they believe in the usefulness of reflective practice in teacher autonomy.

1. *Do you have the habit of recalling what you did in the classroom and evaluate the process of your teaching?*

   In response to the first question, 68% of the teachers mention that they frequently recall the past experience of teaching in the classroom and evaluate the process of teaching. However, 32% of the teachers respond that they sometimes rely on the above-said reflective practice. It means that these teachers do not regularly make use of this practice to enhance teaching learning situation. In other words, a considerable percentage of teachers do not seem to give importance to reflective practice, which is highly beneficial to improve the quality of teaching.

2. *Do you believe that insights gained from the analysis of your past class-room teaching experience help you to ensure effective teaching?*

   While responding to the second question, 82% of the teachers accept that they have gained insights from the analysis of past teaching experience to a greater extent. But 16% of the teachers have selected the second option and they say that they have benefited somewhat from the analysis of past teaching experience. Only 2% of the teachers admit that they have gained very little from reflective practice. On the whole, the responses of the teacher participants reveal that they do not deny the beneficial impact of reflective practice on language teaching.

3. *Do you agree that teachers should stick to a single method of teaching to teach different language aspects?*

   The notion of sticking to a single method of teaching for teaching the different language aspects is not acceptable to majority of the teachers. About 92% of the teachers do not like to depend on a single method of teaching, whereas only 8% of the teachers are in favour of single method of teaching. In other words, most of the teachers want to show interest in eclectic approach for effective language teaching.

4. *Do you think that prescribed textbook alone is sufficient for teaching English to your students?*

   All the participants in the survey clearly mention that textbook alone is not enough for teaching English. It means that the textbook has to be supplemented by teachers using materials from various sources to improve the learning process. It also implies that teachers have to play an important role in selecting apt materials to suit the requirements of their students.

5. *If your answer is 'No' for the fourth question, please tick the materials that you use to supplement the prescribed textbook?*

As far as the materials to supplement the textbook are concerned, the following three sources seem to be highly prominent among teachers – (a) materials from web resources, (b) materials from print and electronic media, and (c) reference books. It is found that 67.3% of the teachers use materials from web resources and that 69.4% of the teachers make use of materials from print and electronic media. Meanwhile, 45% of the teachers use reference books to supplement the prescribed textbook. The responses of the teachers imply that teachers are willing to select the appropriate materials from different resources to make their teaching more effective and interesting.

6. *What are the factors that you take into account while designing language activities for your students?*

From the survey, it is inferred that teachers are highly thoughtful of the factors that should be taken into account while designing language activities for their students. About 66% of the teachers claim that the language aspect to be taught, and teaching aids/facilities should be given high priority. Besides, 46% of the teachers feel that the attitude of the students should be taken into account while designing language activities, and 36% of the teachers think that the allocated time for each class is important while designing language activities. It is clear that the factors such as language aspect to be taught, teaching aids/facilities, attitude of students, and time allotment are given much importance in view of the process of designing language activities.

The overall analysis of the responses of the teachers to the questions in the second part of the questionnaire reveals the mindset of the teachers regarding reflective teaching, methods, and materials. It highlights the following features: (a) Around 68% of teachers frequently make use of reflective practice for professional development; (b) About 98% of the teachers recognize the positive aspects of reflective practice; (c) Almost 92% of the teachers prefer eclectic approach in view of the choice of methods of teaching; (d) All the teachers selected for the study say that prescribed text book is not sufficient for teaching, which needs to be supplemented by teachers; (e) A majority of the teachers are in support of using materials from print and electronic media, web resources, and reference books; and (f) About 66% of the teachers suggest that teaching aids/facilities and language aspect have to be taken into account when designing a language activity.

It can be seen that the responses of the teachers for the first two questions support the first hypothesis framed for the study "Teachers recognize the usefulness of reflective practice in enhancing their quality of language teaching". This shows that they recognize the usefulness of reflective practice in language teaching. Moreover, teachers seem to understand the importance of teacher autonomy in supplementing prescribed textbooks and in designing language activities according to the needs of their students.

Reflective Practice of Teachers

The third section of the questionnaire has questions related to a recent class handled by the participants, and it is concerned with various aspects of the class like language aspect, type of language activity, confidence level of the teacher before and after the class, goal of the lesson, participation of the students, insight from experience, and future plan based on his experience. The analysis of the responses of the teachers may show how far the reflective practice is helpful to ensure the autonomy of the teachers in language class.

The first question is about the language aspect selected for the study. It has been found that about 49% of the teachers have selected speaking as the language aspect for their class. Then 24.5% of the teachers have mentioned that grammar has been the language aspect of their class. It has also been noted that 8.2% of the teachers have responded that listening has been the language aspect of their class. Besides, the remaining teachers have chosen reading, speaking, and writing aspects. Based on the responses of the teachers, it has been inferred that most of the teacher participants have selected speaking skills in view of the top priority given to oral proficiency in workplace contexts.

Teachers have to rely on group work, individual work, and pair work to conduct language activities. While responding to the second question, 69.4% of the teachers have selected group work, 14.3% have chosen pair work, and another 14.3% have selected individual work. It has been noted that group work has been the favourite option chosen by majority of the teachers for conducting language activities in the class. Moreover, the responses of the teachers to the third question have showed that teaching aids, worksheets, and materials from various resources have been extensively used by teachers to enhance the quality of teaching.

Regarding the self-confidence level of teachers before handling the class, it has been observed that 55% have been very confident and the remaining 45% have been somewhat confident. In other words, all the teachers have been positive in their approach towards the class to be handled. However, there has been some variations in view of their satisfaction after their class. About 38.8% have been very satisfied and 59.2% have been satisfied in view of their class handled recently by them. Only 2% of the teachers have been highly dissatisfied with their class. On the whole, it has been clearly evident that majority of the teachers have had satisfaction over the effectiveness of the classroom teaching.

In response to the question on the achievement of the objective or goal of the lesson plan, 49% have achieved the goal to a greater extent in terms of the achievement of the goal of the lesson plan. The achievement of the goal of the lesson plan to a greater extent has been supported by numerous factors. These include proper planning, following the lesson plan, designing the plan focusing on course outcome, using web resources, analysing the students' interest, and engaging in activity-based learning. In addition to this, teachers put forth factors like the attitude of the students towards learning a new concept and learning it in a collaborative classroom. Above all, teachers believe the importance of the rapport with their students, which is also a key factor in achieving the goal of the lesson plan to a greater extent.

Then, 49% have selected the option 'somewhat' and 2% have chosen the option 'very little' in view of the achievement of the goal of the lesson plan. Based on their responses, the following have been identified to be the reasons for not achieving the goal to a greater extent – (a) large class, (b) negative attitude of students, (c) mother tongue medium of instruction in school, (d) unsuitable textbook, and (e) lack of materials to supplement the textbook.

In the case of participation of the students in the classroom activities, 67% of the teachers have responded that their students participated often during the class. The top most reason attributed for the active participation of the students has been the well-planned language activities known for group work and an element of fun. The second important reason has been the appreciation or motivational remarks of teachers to encourage the students to participate in interaction during the language class. The other reasons have been readiness to learn new language items, clarity in instructions, and significance/usefulness of the language activity. However, 25% have selected the option 'occasionally' and 8% have chosen 'rarely' while responding to the question regarding student participation in the class. The major reasons given by the teachers for poor participation of students have been (a) lack of self-confidence, (b) fear of being ridiculed, (c) stage fright, (d) lack of interest, (e) shy behaviour, and (f) low proficiency in English.

Almost all the teachers have responded positively to the open-ended question on the skills/knowledge gained from past classroom experience. According to them, they have acquired the skills/knowledge such as ability to explain the instructions for an activity clearly, rapport with the students, use of motivational remarks and group work. Though they have admitted that they would like to make use of the insights from past experience, they have failed to explain how they would make use of them in future classes.

The last question in the questionnaire has been about the future plan made by the teachers based on insights gained from past classroom experience. Most of them have expressed their willingness to utilize technological aids such as smart board or projector in the classroom to make the class more interesting and interactive. Besides, they want to plan for ICT (Information and Communication Technology)-based teaching and activities using online platforms in a more effective manner. Some of them have expressed their desire to rely on critical thinking skills for effective teaching. However, their action plan has not been clear in terms of the selection of materials, teaching techniques, technological aids, and language activities to improve teaching learning process in language classes.

The analysis of the third section of the questionnaire reveals that teacher autonomy can be achieved through reflective practice. The questionnaire using reflective model shows that they pass through six steps like description, feeling, evaluation, analysis, conclusion, and action plan. When they analyse the success or failure of classroom teaching, they get insights which make them come out with the plan to deal with the future classroom situations. The insights gained from past experience empower them to become autonomous in selecting apt materials and techniques to make their classes more effective. The only limitation is that teachers do not seem to explain the action plan clearly based on the past classroom experience. Apart from this, it can be concluded that reflective practice is helpful to teachers in taking decisions regarding the use of effective teaching methods, materials, and techniques. Therefore, the second hypothesis framed for the study, "Teacher autonomy in language classroom can be ensured when teachers adopt reflective practices for effective teaching" is proved right.

Findings

The study highlights the point that teachers are aware of the uses of reflective teaching practice in language classes. Around 68% of the teachers have the habit of recalling the past experience of teaching to develop their teaching skills. In addition, 98% of the teachers agree with the usefulness of insights gained through reflective practice. As a result of their awareness about the benefits of reflective practice, majority of the teachers believe in eclectic approach and show interest in supplementing prescribed textbook to suit the needs of their students. Apart from the activities given in the textbook, they design activities taking into account the factors like time, language aspect, and the availability of technological facilities. These findings confirm that teachers accept the merits of reflective practice and the necessity of teacher autonomy to ensure effective teaching learning process.

Conclusion

The present study conducted using google form questionnaire based on Gibbs reflective model has attempted to study how far reflective practice has been followed by teachers of English at the tertiary level. Among the six steps of this model, the analysis of the teachers' responses shows that majority of the teachers have not been clear regarding the last two steps like 'conclusion' and 'action plan'. On the basis of the insights gained from past classroom experience, they do not seem to take decisions and implement them in future classroom situations. This is the major limitation that can be overcome by focusing the attention on 'conclusion' and 'action plan'. Moreover, this shows that training programme on reflective teaching with emphasis on above said aspects is imperative to make the teachers more autonomous in the process of language teaching. The feasible modes of reflective teaching will immensely result in professional development of the teachers. Language teachers in the postmodern era have to go beyond the textbook and the classrooms experience. They need to frame their action plan constantly and continuously throughout their lives to achieve the goal of effective teaching. As Larrivee (2000) points out, critical reflection is not simply an approach to teaching but it is a way of life.

References

Al-Ahdal, A. (2014). Reflective teaching and language teacher education programs: A milestone in Yemen and Saudi Arabia. *Journal of Language Teaching and Research, 5*(4), 759–768.

Bartlett, L. (1990). Teacher development through reflective teaching. In J. C. Richards, & D. Nunan (Eds.), *Second Language Teacher Education* (pp. 202–214). Cambridge: Cambridge University Press.

Carper, B. A. (1978). Fundamental patterns of knowing in nursing. *Advances in Nursing Science, 1*, 13–24.

Driscoll, J. (2007). *Practising clinical supervision: A Reflective approach for health-care professionals*. Second Edition. Edinburgh: Bailliere Tindall Elsevier.

Gibbs G. (1988). *Learning by doing: A guide to teaching and learning methods*. London: Further Education Unit.

Impedovo, M. A., & Malik, S. K. (2016). Becoming a reflective in-service teacher: Role of research attitude. *Australian Journal of Teacher Education, 41*(1), 99–113.

Johns, C. (1995). Framing learning through reflection within Carper's fundamental ways of knowing in nursing. *Journal of Advanced Nursing, 22*, 226–234.

Kolb, D. A. (1984). *Experiential learning: Experience as the source of learning and development*. Englewood Cliffs, NJ: Prentice-Hall.

Larrivee, B. (2000). Transforming teaching practice: Becoming the critically reflective teacher. *Reflective Practice, 1*(3), 293–307.

Priya Mathew, P., Mathew, P., Prince, & Peechattu, J. (2017). Reflective practice: A means to teacher development. *Asia Pacific Journal of Contemporary Education and Communication Technology, 3*(1), 126–130.

Robichaux, R. R., & Guarino, A. J. (2012). The impact of implementing a portfolio assessment system on pre-service teachers' daily teaching reflections on improvement, performance and professionalism. *Creative Education, 3*(3), 290–292.

Schön, D. A. (1983). *The reflective practitioner: How professionals think in action*. New York: Basic Books.

Sellars, M. (2012). Teachers and change: The role of reflective practice. *Procedia - Social and Behavioral Sciences, 55*, 461–469.

Smyth, J. (1993). Reflective practice in teacher education. *Australian Journal of Teacher Education, 18*(1), 11–16.

Ur, P. (1999). *A course in language teaching: Practice and theory*. Cambridge: Cambridge University Press.

Xhaferi, B., & Xhaferi, G. (2012). Enhancing learning through reflection – A case study of SEEU. *SEEU Review, 12*(1), 53–68.

Appendix A

Test Items (Pre-Tests)

Part A (9 × 2 = 18 Marks)

1. Write 'Wh' questions to the following:
   a. .... color is the bird?
   b. .... umbrella is this? Is it yours?

2. Add suitable prefixes to the words
   a. ..... logic (without logic)
   b. ...... pure (not pure)
   c. .....sense (without a sense)
   d. .... formal (relaxed and casual)

3. Fill in the blanks with correct preposition:
   a. Vimala goes .... school every morning .....nine.
   b. I begin my day ......a cup of coffee prepared .....my mother.

4. Complete the dialogue with correct 'Yes' or 'No' questions.
   a. Can you speak Arabic?
   ...., but not very well.
   b. Is your brother here?
   ....He is away on business at the moment.

5. Complete the sentences using the given adjectives in the correct degree:
   The Marina beach in Chennai is one of the ....(fine) and the second
   ....(long) beach in the world.

6. Frame 'Yes' or 'No' questions for the following sentences.
   a. I play cricket.
   b. We love ice cream.
   c. Sita is watching TV.
   d. You are a good person.

7. **Fill in the blanks with right articles:**
   a. .... boy with a cap is my cousin.
   b. ....elephant is ..... wild animal.
   c. I love ..... music very much.

8. **Complete the dialogue framing suitable questions:**

   (A patient explains her problem to the doctor)

   Patient Good morning, doctor.
   Doctor Good evening, Please be seated. Tell me ....?
   Patient ....
   Doctor How many days you suffer? Any other discomfort?
   Patient ....
   Doctor Take these tablets for a week. One tablets three times a day.
   Patient ....

Part B (2 × 16 = 32 Marks)

9. Read the following passage and answer the questions:
   Most penguins build their nests on the ground. They carry pebbles and plants from the beach and use them to build their nests. Often, they steal pebbles from each other if they get the chance. Penguins usually lay eggs and both parents crouch over the eggs to keep them warm.

   Emperor penguins lay their eggs in the middle of winter. This is so that the chicks will arrive early in the spring. An emperor penguin lays only one egg. A father penguin takes this in front of his body and rests it on his feet. The chick sits against his warm body until it is old enough to stand the cool.

   Even so, many chicks die of cold before the spring comes. Penguins are good parents while one is looking after the chicks the other brings food. It brings back fish and other small animals and when it has chewed them a little, the chick pushes its head into the parent's mouth to reach for the food later. When chicks are older, all the parents come out to feed together.

   All the chicks stay together in one place where they keep themselves warm and safe from other animals. Big birds will attack small penguins, but they do not often attack them if they are in groups. United they are strong.

Questions:
   1. Where do penguins build their nests?
   2. How do penguins perform their parental duty?
   3. How are the penguin chicks fed?

4. By whom are the small penguins attacked?
5. What is the moral of the passage?
6. Complete the sentences:
    i. All the chicks gather at one place to keep themselves ....
    ii. Many of the chicks die of ....

7. Match the following

1.	Emperor penguins	- brings fish
2.	Father penguins	- attack small ones
3.	Big birds	- pebbles
4.	Nest	- lay eggs

10. **Write a letter to your asking for two days leave. (8 Marks)**
11. **Develop the hints into a readable passage and give a suitable title. (8 Marks)**

    A rich farmer – lot of land – cattle and servants – two sons – happy life – after some years younger son unhappy – asked his share of the property – wouldn't listen to father's advice – got his share – sold them all – went away to another country – fell into bad ways – soon all money gone – poor – no one to help him – understood his mistake.

Appendix B

Test Items for Post-Test

Part A (18 Marks)

1. Identify the countable and uncountable nouns in the following sentences:
   a. Windows are made of glass.
   b. Iron is a metal.

2. **Add suitable prefixes or suffixes:**
   a. Locate: displace
   b. Deforest: clearing of forest
   c. Applicable: cannot be applied
   d. Require: something necessary (ation, in, ment, dis)

3. **Frame "Wh" questions to the responses given.**
   a. I washed my motor cycle last week. (When)
   b. She spends most of her time in watching films. (How)

4. **Fill in the blanks with suitable comparative adjectives:**
   a. The Ganges is .... (long) than Cauvery.
   b. Gold is ....(expensive) than Silver.

5. **Fill in the blanks with correct prepositions:**
   a. Henry Ford was born .....the 30th of July 1863.
   b. ....1903, he founded the Ford Motor Company.
   c. Henry Ford became one ....the richest and best-known people in the world.
   d. Ford's first model T could then be assembled ....just 93 minutes.

6. **Fill in with right articles:**
   a. I want ...... apple from the basket.
   b. ....church on the corner is progressive.
   c. Can I borrow ...... pencil form your pile of pencils and pens?
   d. One of the students said, "...........professor is late today"

7. **Fill in the blanks with correct tense.**
   a. Weather is created by the heat of the sun. when the sun ....
   (Shine) on the earth, the air close to the surface........ (Heat) up. The
   higher it ....(go), the cooler it ....(become)

8. **Complete the dialogue framing suitable questions:**

   A: Hi, ...?
   B: I am fine. ....
   A: It has been long time we met. ....?
   B: I am working in Chennai.
   A: That's nice. I am in Bangalore.
   B: oh good. I will meet you soon. I have to go now. Bye
   A: ....

9 **Choose the correct adverbs and complete the sentence sentences: (Jan – 2018)**
   (a) He writes .... (slow)
   (b) She responded .... When she was interviewed by her supervisor (confident)
   (c) He is laughing ....(loud)
   (d) Ram sings .... (nice)

Part B (2 × 16 = 32 Marks)

10. **Read the following passage and answer the questions given below: (Jan – 2017)**
    Space is a dangerous place, not only because of meteors but also because
    of rays from the sun and other stars. The atmosphere again acts as out
    protective blanket on earth. Light gets through, and this is essential for
    plants to make the food good that we eat. Heat too makes our environment tolerable and some ultraviolet rays penetrate the atmosphere.
    Cosmic rays of various kinds come through the air from outer space,
    but enormous quantities of radiation from the sun are screened off. As
    soon as men leave the atmosphere, they are exposed to this radiation
    but their spacesuits or the walls of their spacecraft, if they are inside,
    do prevent a lot of radiation damage. Radiation is the greatest danger
    to explore in space. Doses of radiation are measured in units called
    "rems". We all receive radiation here on Earth from the sun, from cosmic rays and from radioactive minerals. The 'normal' dose of radiation
    that we receive each year is about 100 millirems (0.1 rem); it varies
    according to where you live, and this is a very rough estimate. Scientists
    have reason to think that a man can put with far more radiation than

this without being damaged; the figure of 60 rems has been agreed. The trouble is that it is extremely difficult to be sure about radiation damage; a person may feel perfectly well, but the cells of any of his internal organs may be damaged and this will not be easily discovered by the person.

Early space probes that radiation varies in different parts of space around the Earth. It also varies in time because, when great spurts of gas shoot out of the sun (Solar flares), they are accompanied by a lot of extra radiation. Some estimates of the amount of radiation in space, based on various measurements and calculations are as low as 10 rems per year, others are as high as 5 rems per hour. Missions to the moon (the Apollo flights) have had to cross the Van Allen belts of high radiation and during the outward and return journeys; the "Apollo 8" crew accumulated a total dose of about 200 millirems per man. It was hoped that there would not be any large solar flares during times of Apollo moon walks because the walls of the lunar excursion modules (LEMS) were not thick enough to protect the men inside, though the command modules did give reasonable protection. So far, no dangerous doses of radiation have been reported, but the Gemini orbits and the Apollo 8 missions have been quite short. We simply do not know yet how men are going to get on when they spend weeks and months outside the protection of the atmosphere, working in a space laboratory or in a base on the moon. Drugs might help to decrease the damage done by radiation, but no really effective ones have been found so far. At presentation, radiation seems to be the greatest physical hazard to space travelers, but it is impossible to say that just how serious the hazard will turn out to be in future.

I. Choose the correct answer for the following questions: (4 × 1 = 4)

1. Scientists have fixed a safety level of
   (a)   10 rems per year
   (b)   60 rems per year
   (c)   100 millirems per year
   (d)   5 rems per hour.

2. The spacemen were worried about solar flares when they were
   (a)   crosrsing the Van Allen belts
   (b)   setting up a moon base
   (c)   exploring the surface of the moon
   (d)   waiting in the command module

3. When men spend long period in space how will they protect themselves?
   (a)   by taking special drugs
   (b)   by wearing special suits

(c)   by using a protective blanket
(d)   no solution has been found yet

4. Which of the following is true?
   (a)   radiation seems to be very harmful
   (b)   radiation can damage internal organs
   (c)   drugs may not help to reduce the damage
   (d)   we receive radiation on earth from the moons

II. Write the statements True or False. (8 × 1 = 8)

   1. The atmosphere screens off the Earth from excessive radiation.
   2. Everyone on earth is exposed to exactly the same amount of radiation.
   3. Solar flares are not dangerous.
   4. Space is a dangerous place because it is not fully explored.
   5. The "Apollo 8" missions have been quite long in duration.
   6. The drugs have found to decrease radiation are ineffective.
   7. The greatest physical hazard to space travelers is remaining for long hours in space.
   8. In space travel, space suits are absolutely necessary for the scientists.

III. Answer the following questions in one or two sentences: (2 × 2 = 4)

   9. What is radiation and how it is measured?
   10. How many missions have crossed the space?

I. Write a letter to your uncle who has presented a book to your birthday.
   (1 × 8 = 8)

II. Write a paragraph on cell phones and its uses in modern technology.
   (1 × 8 = 8)

Index

Page numbers in *italics* refers figures; **bold** refers tables

Abdullah, M.R.T.L. 242–243
Abirami, P.G. 4, 12
activity-based learning 150
Adams, K. 148
Adrian, M. 45
Aebersold, J.A. 206
Agarwal, P. 4, 20
Agnew, L. 236
Ahmad, M.A.K.B.A. 122
Ahuja, G.C. 201, 203
Ahuja, P. 201, 203
Akimove, Aleksander 127
Akturk, A.O. 32
Al-Ahdal, A. 275
Al Burns, S. 16
Alderson, J.C. 96
Alexander, R. 235
Aller, B.M. 145
Alley, M. 145
All India Council for Technical
    Education (AICTE) 11–12, 42, 57,
    119, 132, 258, 264
Alves, A.C. 123
AMA Critical Skills Survey (2012) 23
Ambigaphaty, P. 239
American Association of colleges and
    Universities (AACU) framework
    (2007) 5
American Commission on Teacher
    Education 256
American Library Association, 1989 8
Anagun, S.S. 5, 15
"Anaphora," 137
Anderson, J. 207
Anderson, V.J. 89
Andrews, R. 240
Andria Babu, B. 271–281

Aniswal, A.G. 239
Apriana, A. 86
Aramburo-Lizarraga, J. 15
Arboleya, A. 123
Aryadoust, V. 96
Aspiring Minds 4, 11–12, 162
Aspiring Minds–AMCAT (2014) 56–57
As-Salam College of Engineering and
    Technology 91–93, **92**
Assessment and Teaching of 21st
    Century Skills (ATC21s) (2010) 5
Attewell, P. 43
Attwell, D. 242
audio-lingual method 85
Australian Council for Education
    Research (ACER) 5
Aviation Disaster 126–127
Awang, H. 122

Bachman, L.F. 193–194, 207
Bangalore project 64, 194
Bano, Y. 4
Barnett, M. 204, 207
Barratt, K. 35
Bartlett, L. 271
Basturkmen, H. 147
Bath, D. 56
Battelle for Kids organization 5
'2014 Beas River Tragedy,' 128
Beglar, D. 64
Bell, L.C. 207
Berg, I. 43
Berry, R. 57
"Better English for Better Employment
    Opportunities," 44
Bhattacharya 162
Bialystok, E. 164

Bimmel 29
Blackboard 226–228, 230, 240
Black, R.W. 55
Blake, William 137
blended learning 240
Blom, A. 44
Bloom, B.S. 145
Bloomfield, L. 97
Bloom's taxonomy 7, 15
Bond, G.L. 208
Bondjema 90
Boonkit, K. 164
Boston, T. 273–274
bottom-up theory 203–204
Boyle, O.F. 97, 113
Branden, K.V.D. 70
Braverman, H. 43
Breen, M. 62, **63**
Bridging the Skills Gap, 2012 9, 41, 55
"Bright Star," 135
Brown, A.L. 32
Brown, H.D. 87, 97, 163
Brown, J.S. 237
Bruner, J. 240–241
Bugon, G. 16
Bureau of Industrial Consultancy
    Services (BICS) 57
Burns, S. 16
Busaidi, S. 16
Butterworth, J. 145
Bycina, D. 209
Bygate, M. 163, 182

Cai, Yuzhuo 44
Cambridge First Certificate
    Examination 235
Canale, M. 193
Career Builders annual survey (2017) 41
Carless, D. 64
Caroselli, M. 35
Carper, B.A. 274
Carrell, P.L. 204
Central Institute of English and Foreign
    Languages (CIEFL) 256
Chaiklin, S. 239
chalk-and-talk method 225–226, 229
Chamot, A.U. 29, 205–206
Chang, Y. 202
'Charkhi Dadri Mid-air collision,' 126
Chastain, K. 150
Chaudhary, Shreesh 256
Chaudron, C. 29
Chen, C. 27

Chenoy, D. 44
Chernobyl nuclear disaster 126–127
Chitra, N. 61–78
CII 12
Claghorn, Patricia 55
Clarke, M.A. 207
classroom-based activities 35, 240
Cleanth Brooks' notion 133
Clement, A. 45, 161
Coetzee, S. 13
cognitive strategies **29–31**, 30–31, 33,
    176, 183, 185, 205, 207
Cohen, A. 29, 207
Cole, K. 150
collaboration skills 5–8, 10, 15, 22–23,
    26, 29, **30–33**, 34–36, 35, 55, 122,
    238, 240, 242–243
collaborative learning 26, 120–123, 125
Collie, J. 132
Collins, A. 237
Collins, R. 43
Common European Framework of
    Reference (CEFR) 7, 166–167, **167**
Communicational Teaching Project
    (CTP) 64
communication skills 7, 14, 24–26, **25**,
    119–124, 128, 130, 132, 142,
    145–146, 148, 162, 230–232, 236,
    260, 263, 269; organization's
    productivity 47, **48**, **49**, 49–50;
    remedial measures 53, **53**; studies on
    44–45; task-based language teaching
    see task-based language teaching;
    Tiruchirappalli district, enterprises in
    50–52, **50–52**, 51, 55
communicative competence 62, 132,
    193
communicative language teaching
    (CLT) 62, 86, 266
Computer and Technology Industry
    Association (CompTIA) 41
computer-assisted language learning
    (CALL) 230–231
'conceptually driven' approach 204
conventional teacher-centred
    instruction method 15
conversation analysis (CA) 193
Cooney, E. 23–24
CoreEL Technologies 12
Cornelius-White, J.H.D. 242
corporate disaster 128–129
COVID-19 pandemic 3, 7, 10, 144
Crandall, J. 206

creativity skills 26–28, 27, 28
Crew Resource Development 126
critical thinking skills 4–7, 16, 22–24, 24,
    29, 31–33, 34–36, 56, 123–124, 130,
    144–146, 149, 243, 248, 275, 280
Cropley, D.H. 27, 149

Dadour, E.S. 205
Daly, Shanna 28
Dansereau 29
Daud, Z. 122
Deal, R.M. 45
"Death be not proud," 134–135
Deb, G. 163
declarative knowledge 32
Dede, C. 4, 26
deep-rooted skills deficiencies 43
Delors, J. 4
Deshler, D.D. 29
Devine, J. 32, 204
DeVries, S.H. 241
Dhanavel, S.P. 132–142, 133
Dharini, R.K. 161–186
direct method 85
Donato, R. 237
Donne, John 134, 137
Donnell, J.A. 119, 145
Dornyei, Z. 67, 163, 165, 167–170
'Downfall of Nokia,' 126, 128
Driscoll, J. 273
Driscoll model 273–274
Dubin, F. 209
Dudely-Evans, T. 121
Duke, N. 210
Dwyer, C. 240
Dyatlov, Antoly 127

Economist Intelligence Unit (2015)
    43–44
Eisner, E. 150
e-learning 240–243
Ellis, R. 29, 62–63, 63, 64–65, 65,
    66–71, 74, 166
El Menoufy, A. 95
Emans, R. 208
enGauge framework of Metiri Groups
    and NCREL (2003) 5
engineering curriculum 4, 9, 12–14, 16;
    AMA Critical Skills Survey (2012)
    23; applied linguistics 99; classroom-
    based activities 35; collaboration 26;
    communication 24–26, 25, 34;
    Computer Science and Engineering

students 100; creativity 26–28, 27,
    28; critical thinking 23–24, 24, 34;
    explicit instruction 34; face learning
    98; instruments 99–100; job market
    22–23; learning strategies 28, 28–33,
    29–33; level of efficiency and
    duration 99; listening sub-skills 100,
    102–103; Mechanical Engineering
    students 100; mini-dialogue 105;
    poetry see poetry; reading sub-skills
    106–107; research design 100, 101;
    research questions 99; skills set
    20–21; speaking sub-skills 103–106;
    specific activities 105; 21st Century
    skills 21–22, 22; sub-skills see
    listening, speaking, reading, and
    writing (LSRW) skills; writing
    sub-skills 108–109
English as a Foreign Language (EFL)
    courses 97
English as a Second Language (ESL)
    91–93, 92, 97; see also task-based
    language teaching; TTT method
    91–93, 92
English for Specific Purpose (ESP) 45;
    AICTE 119; Aviation Disaster
    126–127; case study method
    124–125, 130; collaborative learning
    120–121; communication skills
    required in engineering workplace
    119; components 121; corporate
    disaster 128–129; course and
    classroom activities 121; course
    development 121; definition 119;
    discourse analysis 121; first-semester
    English course, NIT-T students 120;
    focal point of communication 120;
    Indian educational department 119;
    needs analysis 121; power plant
    disaster 127–128; project-based
    learning methodology 120–123;
    task-based language learning (TBLL)
    method 121
English Language Proficiency (ELP) 41,
    161; organizational reputation,
    degradation 46–47, 47; skills gap
    studies 42–43
English language skills 14–15; overall
    performance levels after study period
    112, 112; overall performance levels
    before study period 110, 111;
    percentage and performance levels
    110, 110; post-test performance

percentage of participants 112–113, *113*; pre-test performance percentage of all participants *111*, 111–112
Eskey, D. 203–204
Estaire, S. 66
European Commission (2007) framework 5
European Framework for the Digital Competence of Educators (DigCompEdu) 5
explicit knowledge 71
extensive listening 102–103
Ezekiel, Nissim 139

Facebook 225–228, 246
Fadel, C. 6
Faerch, C. 96
Fatmawati, A. 15
Field, J. 96
Field, M.L. 206
Fitriati, S. 177
Fitzpatrick, A. 235
Flavell, John, H. 32
Fleming, P.J.S. 241
focus on form 65, 68, 72, **72**
focus on language 72, **72**, 77
Form, W. 43
Freire, P. 241
Frost, Robert 138
Fuchs, D. 207
Fuchs, L.S. 207

Gallup 13
Gamage, G.H. 30
Ganayem, A. 15
Ganesh, S. 56
Gani, S.A. 183
Garber, P.R. 35
Garner, R. 205
Gaurav, J. 20
Geddes, Marion 234
Geetha, R. 20–36, 237–249
General Motors 129
'General Motors Cobalt Ignition Switch Crisis,' 129
Ghanta, S. 204
Gibbons, P. 241
Gibbs, Graham 272–273
Gibbs reflective cycle 272–273
Gibran, Kahlil 141
Gibson 24
Gilly Salmon's five-stage model: business meeting 244; e-learning 242–243, 246, **246**; flow diagram

242, *243*; persuasive oral presentation 244; process description 244; 'Professional Communication Skills (PCS)' course 244; sample learning tasks **245**; stage 1, 243, 246; stage 2, 243, 247; stage 3, 243, 247; stage 4, 243, 247–248; stage 5, 243, 248; technical aspects 242; technical oral presentation 244
global economy 6, 8, 10–12, 22
Globerson, T. 207
Gloucester County College (GCC) 55
Goh, C.C. 96
Gokak, V.K. 257
Goleman, D. 56
Gol, O. 26, 121
Goodman, K.S. 95, 204
Gough, P.B. 204
Gower, R. 265
Grabe, W. 207
Gradual Immersion Method (GIM) 15
grammar-translation method 84–85
Gray, W.S. 203
Guarino, A.J. 275
Gunashekar, Paul 263
Gurukul system 226

Hadfield, J. 89
Hagley, E. 202
Hall, B.T. 145
Hamilton 26
Hammond, J. 241
Hansraj 150
Harbaugh, A.P. 242
Harishree, C. 3–17, 119–130, 201–220
Harmer, J. 52, 97, 164
Hathaway, G.M. 202
Hatip, F. 70
Haythornthwaite, C. 240
Hedge, T. 242
Helps 23
Hess, N. 140–141
Hidayat, S. 122
Hiltrud Dave Eve, P. 83–93
Hiltz, S. 240
Hind 56
Hodgson, I. 56
Hopkins, Gerard Manley 136
Hoshino, Y. 96
Hosp, M.K. 207
Hsu, K. 27
Hunt, A. 64
Hussin, Z. 246
Hymes, D. 193

Impedovo, M.A. 274
Indian Confederation of Industries 42
Indian ESL classrooms: background
  knowledge 190–191; conversation
  analysis (CA) 193; language
  acquisition 189; language teachers
  190; material selection 190–191;
  primary and secondary level
  education 190; recruitment selection
  processes 191; speaking tasks
  194–195; task-based language
  teaching (TBLT) 194; teaching
  speaking 191–192; testing speaking
  192; theory of communicative
  language 193–194; theory of practice
  192–193
India Skills Report (2017) 42
India Skills Report (2021) 20–21
The India Study Channel (2013) 56
Indra, C.T. 133
Industry 4.0 144
information, communications, and
  technology (ICT) literacy 8–9, 15–16
information literacy 8
inquiry-based learning 150
In-Service Education and Training
  (INSET) programmes: cluster
  sampling technique 260; collegiate
  education in Tamil Nadu 257–259;
  curriculum design and materials
  development 268; description 255;
  genesis, in India 256–257;
  implementation guidelines 263–264;
  lack of teaching competence 261;
  language for technical
  communication 267; language
  teaching methodology 266–267;
  module designing 264–265;
  strengthening classroom practice
  268; supervised teaching practice
  261; teachers' preference 261;
  teacher trainers view 262–263;
  teacher-training syllabuses 265–266
INSET programmes see In-Service
  Education and Training (INSET)
  programmes
instrumental motivation 84, 175
integrative motivation 84
intensive listening 98, 100, 102
An Intermediate Course in Listening
  Skills 234
International Association for the
  Evaluation of Educational
  Achievement (IEA) 3

International Business English 235
International Reading Association
  (IRA) 202
International Society for Technology in
  Education (ISTE) (2007) ICT Skills 5
Interpersonal Communication Skills
  (IPCS) 133
Islamiyah, M. 86
Ivleva, N.V. 24, 146

Jacobs, J. 206
Jaelani, Z. 163
Jain, P.M. 89
Janakiraman, V. 263
Jaramillo, A. 206
Jawhar, M. 239
Jayachandran, J. 161
Jayamani, T.S. 145
Jenkins, J.R. 207
Johns, A.M. 121
Johns model of reflection 274
Jones, K.S. 240
Jones, L. 235

Kalton, G. 260
Kampen, M. 3
Karzunina, D. 45
Kassem, M.A.M. 162
Katz, L.F. 43
Kaur, A. 12, 15
Kavita, K.M. 56
Keane 24
Keats, John 135
Kedrowicz, A.A. 145
Keep, E. 43
Kelch, K. 124
Kember, D. 56
Kennedy, S. 96
Khan, S. 163
Kim, M.M.J. 32
Kirschner, P.A. 122
Klimova, B.F. 121
knowledge economy 4, 9, 162
Kolb, D.A. 272
Kolb reflective cycle 272
Kothari Commission 256
Kothari, C.R. 166
Kozma, R.B. 242
Krahn, H. 43
Krashen 89
Kuhn, M. 208
Kumaran, S. 132–142
Kumaravadivelu, B. 66
Kumari, M.J. 214

Kumar, M. 12
Kupper 29
Kustati, M. 184

laboratory-assisted language learning
    (LALL) 230–232
Lake, W. 85
Lamb, C. 242
language learning strategies (LLS):
    cognitive reading strategies 207;
    definition 205; direct and indirect
    strategies 205; metacognitive reading
    strategies 207; reading strategies
    205–206; second-language
    acquisition (SLA) 205; strategic
    reading 206–207
language pedagogy 73
Lan, L.S. 164
Lan, V.T.N. 170
Las-Heras, F. 123
Lavrysh, Y. 121
Lavy, I. 144
Lazar 142
Lazarton, A. 193
learning and innovation skills:
    collaboration 7–8; communication 7;
    creativity and innovation 6; critical
    thinking skills 7; problem-solving
    skills 7
learning strategies 28; autonomy
    28–29; cognition 29–31, 30–31;
    definition 29; metacognition 29,
    32–33, 32–33
Leaver, B. 205
Lee, J.F. 66–67
Lenard, B.P. 25
Lenz 29
lesson plan: benefits 70–71; designs 66;
    Ellis model 67, 67–68; principles
    69–70; Willis model 68, 68–69
Leung, D. 56
lexical approach 87
Liansari, V. 185
Lichtenegger, B. 9, 14
life and career skills: flexibility and
    adaptability 9–10; initiative and
    self-direction 10; leadership and
    responsibility 11; productivity and
    accountability 10–11; social and
    cross-cultural skill 10
Linvill, N. 25
listening skill 95, 139–140, 230–232,
    234–236, 261, 267

listening, speaking, reading, and writing
    (LSRW) skills 14, 46, 56, 72–73;
    see also Test-Teach-Test (TTT)
    method; advocate language skills
    113; common sub-skills strategies 98;
    in engineering curriculum see
    engineering curriculum, sub-skills;
    English as a Foreign Language (EFL)
    97; English as a Second Language
    (ESL) 97; English language skills see
    English language skills; individual
    language skills 97; language learning
    strategies 97; language teaching 96;
    listening skill 95; natural language
    skills 97; pervasive approaches 114;
    productive skills (SW) 97;
    pronunciation 96; reading skill 95;
    receptive skills (LR) 97; research
    design 100, 114; skill orientation and
    rote memorization 97; speaking skill
    95; writing skill 96
literacy skills 3, 8–9
Littlejohn, A. 62
Littlewood, W. 64, 151
Liu, Z. 27
LLC5510MKII 233
L2 learning 63–64
Lochana, M. 163
Long, M. 62, 63, 65, 65
Lowe, G.S. 43
Lucas, G.R.I. 170
Lundsteen, S.W. 95
Lunt 23

Maarof, N. 173
Macready, G.B. 205
macro skills of language see listening,
    speaking, reading, and writing
    (LSRW) skills
Madhavaiah, U. 145
Malihah, N. 73
Malik, S.K. 274
Mallett-Hamer, B. 44
Malupa-Kim, M. 124
Manjula Bashini, K. 255–269
Manpower Group's 2011 42
Mansukhbhai Prajapati's Mitticool
    project 153–154
Marra, Rose 26
Martin 27
Martina, S. 171
mathematics instructional model 15
Mathew, P. 275

Mayer 29
Mayhew, K. 43
McConnell, D. 241
McLean, P. 5
Means, B. 150
media literacy 8–9
Medium and small manufacturing enterprises (MSMEs) 55
Meenakshi Sundaram, C. 259–260, 263
Mekala, S. 21, 28, 119–130
Mercer, C.W. 241
metacognitive experience 32
metacognitive knowledge 32
metacognitive strategies 29, 32–33, 32–34, 205–208, 218
Miceli 25
Millan, H.L. 133
Milton, John 138
Mistar, J. 170, 172
Mohammadipour, M. 164
Mohan, B. 97, 113
Mohd Khairul, A.M.D. 246
Morimoto, C. 123
Moser, C.A. 260
Moss, P. 43
Motivational Strategies in the Language Classroom 168
Mullins, G. 148
Murphy, J. 64
Murphy, K.M. 43
Murugavel, T. 45, 161
Myers, M. 206

Nae, N. 124
Nafalski, A. 26, 122
Nagano, P. 207
Nageswari, R. 145
Naiman 29
Nakano, T.C. 27
Nandhini, R. 95–115, 189–200
National Academy of Science and Engineering 123
National Association of Colleges and Employers (NACE) 42
National Curriculum Framework 9
National Education Policy 56, 255
National Eligibility Test (NET) 258
National Employability Enhancement Mission (NEEM) 57
National Employability Report (2019) 42
National Institute of Technical Teachers Training Institute (NITTR) 258–259

National Institute of Technology (NIT) 258
National Institute of Technology, Tiruchirappalli (NIT-T) 57, 120, 125, 151
National Skill Development Corporation (NSDC) 11–12, 43, 45
National Workforce Literacy Project (2010) 44
Nayak, S. 161
Nazarieh, M. 32
Near Future Scenario 225–226
Nestsiarovich, K. 25
Newman, S.E. 237
"Night of the Scorpion," 139, 141
Nithyanand, S. 4
Nonaka, I. 248
Norlidah, A. 246
notional-functional approach 16
Nunan, D. 62–63, 63, 64–66, 69, 121, 147, 149, 242
Nutman, P.N.S. 25
Nuttall, C. 209

OCTs *see* oral communicative tasks
old chalk-and-talk method 226
Olsen, L. 206
O'Mallety, J.M. 205–206
O'Malley Michael, J. 29
online meeting portals 225
Oprandy, R. 95
oral communicative tasks (OCTs): core tasks 166; initial tasks 166; motivational strategies 163, 165; self-introduction 166; speaking tasks 164; supporting tasks 166–167; TBLT program 164
Osterholt, D.A. 35
Oxford Economics (2012) 55–56
Oxford, R. 29, 97–98, 113, 165, 205–207
"Ozymandias," 137

Palmer, A.S. 193–194
Panda, R.C. 162
Pandey, M. 44
Pandey, P. 44
Panitz 26
Pardede, P. 23, 26
parental supervision 84
Paris, S. 206
Paris, S.G. 206

Partnership for 21st Century Skills
(P21) 13–14, 16, 21–22, 22; Battelle
for Kids 5; information, media, and
technology skills 8–9; learning and
innovation skills 6–8; life and career
skills 9–11
Parvaiz, G.S. 44–45
Passerini, K. 240
Patel, M.F. 89
Patnayak, K.K. 162
Patyk, G. 208
Paudel, P. 88, 90, 93
Payne, J. 43
Pazos, Pilar 26
Pearson Foundation 13
Pearson group 119
pedagogical strategies (PSs) 162, 168,
168; 5-point Likert scale values 169;
'F' value 170–177, 179–180; learning
strategies 169, 181; L2 oral
proficiency 169, 174; PS-I 170, 171;
PS-II 171, 172; PS-III 172, 173; PS-IV
173, 174; PS-V 174, 175; PS-VI 175,
176; PS-VII 176, 177; PS-VIII 177,
178; PS-IX 178–179, 179; PS-X 179,
180; PS-XI 180, 181; research
question 1, 184; research question 2,
184–185; self-improvement and
affective strategies 179
PeopleStrong 12
Peregoy, S.F. 97, 113
Perfetti, C.A. 207
Perkins, K. 207
Peyton, J.K. 206
Phillipson, R. 96
Piirto, J. 148
Pincas, A. 241
Pintaric, L. 25
Plucker 26
poetry: All India Council for Technical
Education (AICTE) document 132;
for engineers 133; English 133;
humanities 133; Interpersonal
Communication Skills (IPCS) 133;
language 132–133; listening skills
139; reading skills 140–141; sample
poems and teaching points 134–139;
speaking skills 139–140; writing
skills 141–142
Ponmani, M. 237–249
Poorvadevi, D. 189–200
post-test 91–93, 92, 110, 112–115,
286–289

power plant disaster 127–128
Prabhu, N. 64, 66
Prahalad 27
Prameela Priadersini, B.S. 201–220
pre-test 91–92, 92, 93, 110–112, 111,
114, 283–285
problem-based learning 123–124, 150
problem-solving skills 4, 6–7, 15, 22,
29–30, 42, 47, 70–71, 87, 121–122
productive skills see speaking skill
writing skill
profession-oriented courses 146
project-based learning methodology
15, 120–122, 125, 150–151;
communication lapses 125;
definition 122; vs. problem-based
learning 123
PSs see pedagogical strategies
Putri, S.U. 122

Raftopoulos, M. 13
Ragupathi, K. 22
Rahman, N.A.A. 173
Rahmat, M. 44
Rainsbury, E. 56
Rajasekar 202
Ramanakumar, K.P.V. 12
Ramanan, L. 12
Ramaraju, S. 133
Ramaswamy 27
Ram, M.V.R. 145
Rampillon 29
Rani, S.M. 161
Rashid, S.M. 164
Ravichandran, M. 4, 12
Ravichandran, T. 225–236
Ravikumar, B. 145
Ravi, S. 167, 170
Rayner 30
reading proficiency 208; bottom-up
theory 203–204; comprehension
209–210; comprehension scores 219;
description 201; efficient reading
208–209; ESL reading theory and
practice 204; interactive model 204;
International Reading Association
(IRA) report 201; language learning
strategies (LLS) 205–208; limitations
and suggestions 219–220;
metacognitive awareness 207;
metacognitive strategies 218; oral
interaction 212; participants 211;
purpose of reading 202–203;

questionnaires 211; reader and text 201; reading speed on comprehension level 209, **214**, 214–216, **216**, 218; real reading 201; retention level 210, 216–217, **217**; schema theory 204; self-assessment and actual performance 213, **213**; self-learning figures 212; sight vocabulary 218; testing procedure 212; top-down theory 204; verbal think-aloud protocols 207; visual skills 201; visual symbols 202

reading skill 93, 95, 107, 112, 140–141, 202, 206, 211, 219

receptive skills *see* listening skill reading skill

Reddy, B.V.R.M. 11

Redecker, C. 5

reflective teaching: action plan 280; definition 271; Driscoll model 273–274; experience-wise classification of teachers 276, **276**; Gibbs reflective cycle 272–273; hypotheses 275; Johns model of reflection 274; Kolb reflective cycle 272; objective/goal of lesson plan 279; personal profile 276; practice-based professional learning settings 271; pre-service teachers 275; reflective habit 275; Schön model 273; self-confidence level of teachers 279; skills/knowledge 280; views of teacher participants 276–278; well-planned language activities 279

Reppen, R. 96

Richard-Amato, P.A. 237

Richards, J.C. 64, 88, 96–97, 147, 194

Richards, L.G. 27

Riding 30

Riemer, M.J. 24–25

Ripper, M. 148

Rivers, W. 95

Robbins, J. 205

Robichaux, R.R. 275

Roblin 21

Rodgers, T. 194

Rodgers, T.S. 64, 97

Rogoff, B. 239

Roulston, J.D. 55

Rozakis, L. 35

R's approach of paraphrasing 109

Rubin, J. 29

Rubin, L. 255

Ruddell, M.A. 207

Rudman, H.C. 88

Rumelhart, D.E. 204

Russo 29

Rutter, R. 133

Saedah, S. 242–243, 246

Saeki, J. 44

Safari, M. 177

Sahin, I. 32

Salikin, H. 184

Salmon, G. 240–243

Samavedham, L. 22

sample poems and teaching points: "Anaphora," 137; "Bright Star," 135–136; "Death be not proud," 134–135; interpersonal skills 137; "No Man is an Island," 137; "On His Blindness," 138; "O thou my friend," 136; "Ozymandias," 137; "A Poison Tree," 137; rhetorical questions and metaphors 134; "The Road Not Taken," 138; summer days 134; "Thou art indeed just, Lord," 136; "Upon Westminster Bridge," 135

Samuda, V. 66

Sanabria, J.C. 15

Sanchez, M.A.A. 97

Saville-Troike 62

Schön, D.A. 273

Schön model 272–273

Schonwetter, D.J. 27

Schumaker, J.B. 29

Schutz, R. 89

second-language acquisition (SLA) 62, 72, 145, 205

Sellars, M. 274

Sen, Julu 263

Sethuraman, M. 165, 167, 170

Setiyadi, Ag. B. 30

Shabir, M. 185

Shakespeare, William 134

Shanmuga Priya, C. 41–57

Shantha, S. 161–186

Sharabany, R. 207

Sharma, M. 4

Sharma, P. 56

Shawer, Saad 30

Shearer, B.A. 207

Shelley, P.B. 137

Sheth, T.D. 145

Shintani, N. 62

Sibat, M.P. 239

Simkins, M. 150

Singh, P. 44

Sivanganam, L. 44
Sivaraman, K.N. 259, 262
Skehan, P. 62–63, **63**, 65, **65**, 66, 76
"Skills Expectation-Performance Gap: A Study of Pakistan's Accounting Education," 45
skills gap 4, 9, 11–14, 16, 20, 23, 119–120, 148, 162; communication skills 44–45, 47–50, **48**, **49**, 53, **53**; definition 41; low ELP 46–47, **47**, 53–54; recommendations 55–57; research design 45–46; skills mismatch 43–44; studies on 42–43; Tiruchirappalli district, enterprises in 50–52, **50–52**, **51**, 55
Slater, S. 132
small and medium enterprises (SMEs) 44
Smith, F. 204
Smith, K. 55
Smyth, J. 275
Social media 25
soft skills 56
soft skills development 144
Soule, H. 144
Soundiraraj, S. 271–281
"Sounds of Silence," 233
speaking proficiency 73; Common European Framework of Reference (CEFR) 166; English Language Proficiency (ELP) 161; interpersonal strategies 183; lack of motivation 162; language features 164; and language learning strategies (LLS) 165; and learning strategies 182, **182**, 183; mental/social processing 164; motivational strategies in the language classroom 168; motivational strategies of Dornyei 163; observation sheets 167, **167**; one-way ANOVA 170; oral communicative tasks (OCTs) 163–167; participants 166; pre-study questionnaire 166; principles for designing speaking activities 163; PSs *see* pedagogical strategies; psychological barriers 163; resource-based strategies 183; self-improvement strategies 183; SPSS software 170; task-based language teaching (TBLT) 163; technical and professional skills 162; Technical English I & II in engineering courses 161

speaking skill 73, 85, 95, 104–105, 111–112, 114, 139–140, 161–162, 164, 166, 170, 183, 191–192, 239, 279
Specimen listening comprehension test 234
Srividhya, M.A. 12
Stanovich, K.E. 203, 207
State Level Eligibility Test (SLET) 258
Stern 29
Stewner-Manzanares 29
strategic reading 206–207
Strevens, P. 261
suggestopedia 86
Sundaram, Meenakshi 263
Su, Y.C. 97
Swain, M. 193
Swan, M. 64

Takeuchi, H. 248
Tamil Nadu Skill Development Corporation (TNSDC, 2019) 56
Tan, Z. 70
target language (TL) 89–90
task-based language learning (TBLL) 61, 87, 121; *see also* task-based language teaching
task-based language teaching (TBLT) 61, 163–165; assessment 76–77; benefits 70–71; characteristics 64–65, **65–66**; communicative competence 62; definition 62–64, **63**; learners' role 74; lesson plan 66–69, **67–68**; need for 72–73; observation 74; picture description, reasoning-gap activity of 75–76; presentation, practice, and performance 62; results 77; in second language teaching 64; self introduction 74–75; short presentation 76; speaking proficiency 73; student questionnaire 74; students' role 70; task of getting and giving directions 75; task types 71–72; teacher's role 69–70; TED Talk 76
task-based pedagogy 65, **66**
task listening 101–102
Tavalin, F. 150
Taylor 27
Taylor's hierarchy of creativity 27, **27**
Teaching English as a Foreign Language (TEFL) 259
Teaching English as a Second Language (TESL) 258–259, 268–269

teaching speaking *see* teaching
   speaking, Indian ESL classrooms
teaching speaking, Indian ESL
   classrooms: amendments to courses
   191; Common European Framework
   for Reference (CEFR) rating scale
   199, *199*; language ability 191;
   listening activity 192; oral
   presentation 196–197; picture
   description task 195–196, *196*;
   read-aloud tasks 195; role-play task
   197, **197**; task development cycle
   197–198; task types 194; teaching
   and learning pedagogies 191
technical education *see* skills gap
techno-mediated teaching: assignments
   229; Cambridge First Certificate
   Examination 235; communication
   skills 232; computer-assisted
   language learning (CALL) 230–231;
   cybertechnology 227; Google Doc
   228; Gurukul system 226; hearing
   233; International Business English
   235; laboratory-assisted language
   learning (LALL) 230–232; language
   lab 234; language lab facility 233;
   listening 232–233; macro-planning
   and micro-execution 230; Massive
   Online Open Courses (MOOC) 229;
   Near Future Scenario 225–226;
   new-age gadgets and applications
   227; online course platforms 229;
   PowerPoint (PPT) and/or Prezi 228;
   quiz questions 230; supportive tools
   228; techno-push 227; three P's of
   preparation 234; video programs
   235; web-enabled e-weapons 226
'Tenerife Airport Disaster,' 126
Tep, P. 27
Test-Teach-Test (TTT) method:
   advantages 90–91; audio-lingual
   method 85; communicative language
   teaching 86; data analysis 91;
   descriptive analysis 91–92, **92**; direct
   method 85; grammar-translation
   method 84–85; language input
   89–90; lexical approach 87;
   literature review 87–89; one-way
   ANOVA **92**, 92–93; participants 91;
   rural students 83–84; silent way
   method 86–87; structural approach
   85–86; students' prior knowledge 89;
   suggestopedia 86; task-based

language learning 87; total physical
   response 86; urban students 83–84
Thanghun, K. 164
thinking skills: brainstorming session
   152; classroom activities 145;
   communication skills 145–146;
   computer-controlled orientation 154;
   content generation process 145;
   content selection 149; creative
   capacity 148; creative thinking 145,
   149; creativity skill 145; critical
   thinking 145, 149; experimental
   techniques in farming 153; Giant
   Mirrors of Viganella 154; lateral
   thinking 145, 151; learner-centred
   curriculum 146–147; Madhavan's
   innovative approach 153;
   Mansukhbhai Prajapati's Mitticool
   project 153–154; Models of Service
   Excellence 155; needs analysis
   147–149; parallel thinking 145;
   profession-oriented courses 146;
   project-based learning 150–151;
   psychological inquiry 145; risk-
   taking mentality 153; second
   language acquisition 145; Tata Nano
   and Mumbai Dabbawallas 155
Thinking, Technology, Communication,
   and Confidence (2T2C) model 15
Thwaites, G. 145
Tilly, C. 43
Tinio, V.L. 240
Tinker, M.A. 208
Tneh, D. 239
top-down theory 204
total physical response (TPR) 86
traditional form-focused pedagogy
   65, **66**
Tran, M. 145
Trilling, B. 6
Trofimovich, P. 96
Tsai, Y. 202
Tuzlukova, V. 16
21st Century skills 28; autonomy
   28–29; cognition **29–31**, 30–31;
   collaboration 26; communication
   24–26, **25**; competencies for
   employability 13; creativity 26–28,
   27, 28; critical thinking 23–24, **24**;
   definition 3, 21, 29; engineering
   education 11–12; in English
   classroom 14–15; frameworks 4–5,
   13; industry-expected skills 4;

massive employment degeneration 4; metacognition **29**, 32–33, **32–33**; methods and approaches 15–16; overview 16–17; partnership for 5–11; P21 framework 21–22, 22; skills gap 4, 12–14; students' skills set 4, 13

University Grants Commission (UGC) 258
Unni, J. 12
"Upon Westminster Bridge," 135
Ur, P. 72, 271
US Department of Labor (1991) 43
Utama, F. 185

Vandergrift, L. 95
Van Lier, L. 241
Vasanthan, R. 95–115
Vasantha, S. 4
Verma 162
Vignali, G. 56
Vijayakumari, D.G. 12
Vila, C. 26
Visser, D. 13
Vlčkova, Kateřina 29
Vogler 26
Vogt, M.E. 207
Voogt 21
Vygotsky, L.S. 237–241

Wael, A. 168, 176
Wagner, T. 5
Wagoner, S. 205
Walvoord, B.E. 89
Warner, S. 12, 15
Weaver 29
web-enabled e-weapons 226
Wechsler, S.M. 27
Weinstein, E. 29, 207
Weir, C.J. 195
Wenden 29–30
Wheebox 12
White, C.J. 30, 32
Widdowson, H. 204
Widyantero, A. 88–89

Wikipedia 226
Willis, D. 69–71
Willis, J. 62–63, **63**, 66, 68–70, 74, 76
Wilona, A. 173
Wilson, W.J. 43
Wingard, P.G. 255
Woodrow, L. 150–151
Woodward 90
Wordsworth, William 135
World Bank Enterprise Survey 42
World Economic Forum (2015) 5, 55
World Economic Forum's Global Talent Mobility report 42
writing skill 46, 91, 96–97, 112, 114, 141–142, 164

Xhaferi, B. 275
Xhaferi, G. 275
Xu, Jun 23

Yadin, A. 144
Yates, C. St. J. 235
Yoakam, G.A. 210
Yousif, A.S.B. 164
YouTube 226
Yunus, M.M. 122
Yuzhuo, Cai 44

Zander, A. 148
Zanon, J. 66
Zhang, Y. 164
Zidan, W. 15
zone of proximal development (ZPD): apprenticeship and scaffolding 237, 241; collaboration 238; collaborative projects and investigations 242; definition 237; employability skills 239; facilitator role in learner-centred classroom 241; flow diagram 237–238; Gilly Salmon's five-stage model 242–248; goal of 237; incremental development 238; information and communication technology (ICT) 240–242; intended lesson 238; job-seeking process 239; 'learner-centred' approaches 242